MW01056774

BODY BY DARWIN

How Evolution Shapes Our Health and Transforms Medicine

JEREMY TAYLOR

THE UNIVERSITY OF CHICAGO PRESS
Chicago and London

Jeremy Taylor was previously a senior producer and director for BBC Television, and he has made numerous science films for the Discovery Channel and the Learning Channel, among others. He is also the author of *Not a Chimp: The Hunt to Find the Genes That Make Us Human*. He lives in London.

The University of Chicago Press, Chicago 60637
The University of Chicago Press, Ltd., London

Printed in the United States of America

24 23 22 21 20 19 18 17 16 15   1 2 3 4 5

ISBN-13: 978-0-226-05988-4 (cloth)
ISBN-13: 978-0-226-05991-4 (e-book)
DOI: 10.7208/chicago/9780226059914.001.0001

Library of Congress Cataloging-in-Publication Data
Taylor, Jeremy, 1946– author.
Body by Darwin: how evolution shapes our health
and transforms medicine / Jeremy Taylor.
pages; cm
Includes bibliographical references.
ISBN 978-0-226-05988-4 (cloth: alk. paper)—
ISBN 978-0-226-05991-4 (e-book) 1. Evolution (Biology)
2. Human evolution. 3. Evolutionary genetics. I. Title.
QH366.2.T374 2015
576.8—dc23   2015019942

⊗ This paper meets the requirements of ANSI/NISO Z39.48-1992 (Permanence of Paper).

For Linus and Barbara

CONTENTS

INTRODUCTION

Why can't we live forever? Why can't we make human disease a thing of the past? Why is it taking such a long time to cure cancer? These are the sorts of questions that schoolkids often pose to popular science bloggers, student forums, and the "Ask a Scientist" columns of daily newspapers, but they are no less interesting for that. Yet human life expectancy *is* increasing rapidly in countries worldwide and in some now exceeds eighty years. A recent study shows that the difference today in the decrease in human mortality between hunter-gathers and modern Western-lifestyle populations is greater than that between hunter-gatherers and wild chimpanzees. The bulk of this mortality reduction has occurred in just the last four generations of the roughly eight thousand generations of humans that have ever lived. You only have to look at the incredible advances over the last century in surgery, pharmacology, public health, immunology, and transplantation to see what a success story modern medicine has been.

But sweeping statistics like this disguise a perplexing and worrying amount of disease, the incidence of which is getting worse not better. The pattern of human illness, the sickness landscape, is forever changing. To the deceptively naive schoolkid questions above, we can easily add: Why do so many of us suffer from autoimmune diseases such as rheumatoid arthritis, multiple sclerosis, type 1 diabetes, and inflammatory bowel disease? Why are so many people dogged by allergies like eczema and asthma? Why are there epidemics of heart disease? Why are our eyes so peculiarly vulnerable to retinitis pigmentosa and wet macular degeneration? Why are we plagued with backaches, hernias, slipped disks, and dodgy hip joints? If the appendix is a useless fossil, why hasn't it disappeared, saving many of us from appendicitis? Why do women suffer from infertility problems and preeclampsia? Why is there so much mental illness about? And why do so many of us descend into the mental twilight world of Alzheimer's disease in later life?

The medical profession tends to view the human body as a well-designed machine, although it is a machine that has a tendency to break down. It needs constant servicing, and occasionally develops faults that have to be repaired in the same way that a skilled team of mechanics can keep a racing car engine purring sweetly. Medical students are taught that their job as doctors is to fix the machine and keep it running for as long as possible. But the human body is not a machine. It is a bundle of living material that is the product of evolution by natural selection, just like all life on Earth, and there are fundamental differences between machines, or any architectural or engineering structure, and human bodies.

When we commission an architect or engineer to design something in the material world—let's say a new high-rise office building—their first question will be "What's my brief?" We might reply that it must have a certain height, that it must take its energy requirements from solar panels, that we would like the elevators to run up the exterior of one side, that it must otherwise blend in with surrounding architecture, and that it must be capable of lasting two hundred years—whatever. This will result in a blueprint that is rigidly adhered to. If unforeseen problems arise, the architect can go back to the drawing board and start again on one component or another.

The brief for evolution is quite different. Human bodies have design criteria unlike anything you will ever meet in the worlds of architecture and technology. Evolution is not "interested" in health or happiness, or longevity. To put it into Darwinian jargon, it is only interested in maximizing an individual's reproductive fitness. This means fashioning changes in organisms that allow them to adapt to environmental change and reproduce. To the extent that genetic change results in the differential reproductive success of any members of a species, the genes they carry should increase in frequency in a population. Thus, while genes may be immortal, bodies were never intended to be. Evolution will only select for traits in individuals that allow them to survive beyond reproductive age if they improve the chances of survival of the proportion of their genes that occur in sons and daughters, close relatives, and grandchildren. Furthermore, unlike any reputable architect, evolution is blind and witless. It has no blueprint, it cannot look into the future and design accordingly, and there is no sense in which evolution can "see" the nature of a problem and build the perfect answer to it. It has to make do with what is in front of it at any one point in any organism's evolutionary history. It cannot unpick design and structure and go back to basics just because some new selection pressure comes along and requires a corresponding change in structure or function in order for individuals of that species to survive.

Thus engineering and machinery analogies are profoundly misleading and do not help us toward a deep understanding of why we are so vulnerable to disease and degeneration. This is why four pioneers of evolutionary medicine—Randolph Nesse, Stephen Stearns, Diddahally Govindaraju, and Peter Ellison—have recently sought to overturn the engineering analogy once and for all because it is still very entrenched within the medical profession. Because evolution works to maximize reproduction, not health, these scientists say, organisms are bundles of compromises that are full of unavoidable trade-offs and constraints. Second, because biological evolution is so much slower than cultural change, much disease arises from the mismatch of our bodies to modern environments. Also, pathogens can evolve much faster than we can, making sure that infection will always afflict us. Finally, the idea that much human disease is caused by the inheritance of a few defective genes is usually incorrect. The norm is that many gene variants interact with one another and the environment to give rise to disease. Thus, disease and illness will always be a fact of life and difficult to prevent.

Evolutionary medicine allows us to look at the human body in a completely different way and often throws up insights that run counter to prevailing wisdom about what illness is. A very simple, everyday example is the role of fever in infections. When we get the flu, we run a temperature and we often feel depressed and unenthusiastic at the prospect of going about our daily lives. Much over-the-counter medicine is devoted to reducing these febrile symptoms. But since pathogens prefer a temperature that is lower than that of the human body, fever is in fact a sophisticated evolved mechanism to raise body temperature and make the human body as hostile an environment as possible to invading microorganisms.

Peter Gluckman, from the University of Auckland, gives a more complex example by explaining how evolution can throw light on the reasons why breast and ovarian cancer has been on the rise for decades and why breast cancer is now one of the top five causes of death in women in developed countries. Evidence suggests, says Gluckman, that coming late into first menstruation but quickly having a first baby, followed by having a relatively large number of pregnancies all with prolonged lactation, followed in turn by early menopause are all protective against getting breast cancer. This would have been typical of women in the Paleolithic period of our prehistory. However, modern women are the opposite. They menstruate earlier; there is, typically, a large gap between menarche and first pregnancy (which means many menstrual cycles); low numbers of children and short, if any, lactation. Modern women will experience around five hundred ovulations during their fertile life, and it is thought that mechanical injury to the skin of cells covering the ovary, together with high local fluctuations in sex hormones, increases the risk of ovarian cancer. This may explain, suggests Gluckman, why using oral contraceptives, cutting down the number of menstrual cycles a woman endures, seems related to a decreased risk. Allied to this, immature ductal breast tissue in women who have never given birth (full maturation depends on first pregnancy) and the constant regeneration of epithelial cells in the breast urged by cycling estrogen and progesterone, in the absence of prolonged amenorrhea during multiple pregnancies, leads to higher rates of breast cancer. Lack or reduction of breast-feeding also reduces the beneficial washout of pre-cancerous cells into mother's milk.

Thus, women today are mismatched for reproductive biology thanks to a dramatic shift in reproductive behavior involving the use of contraception and hormone replacement therapy, lack of children, fewer children, reduced lactation, earlier menarche, and later menopause. They have a longer fertile life typified by many more wildly fluctuating hormonal states during their many menstrual cycles. But how can we explain the existence of the cancer susceptibility genes—BRCA1 and BRCA2? Specific variants of these genes lose their power to suppress tumor development in breast epithelial cells. Gluckman points out that while most women experience breast cancer in later age, a significant number experience cancer when they are younger. One would therefore expect that the variants of these genes that dramatically increase the risk of cancer would have been selected against such that their frequency in modern populations would have been reduced. But they have not. This suggests to Gluckman that these genes confer some beneficial effects early on in life that balance the deleterious effects that occur later on. The phenomenon is called antagonistic pleiotropy, and it crops up frequently in evolutionary models of human disease. A recent study, Gluckman points out, has shown that carriers of the BRCA1/2 mutations have both significantly increased fecundity *and* post-reproductive mortality. Evolution may have traded off increased fertility earlier in reproductive life for increased risk of death through breast cancer later on.

Given the pleasing and incredibly useful explanatory power of evolutionary thinking like this, you would have thought evolution would feature much more prominently in medical explanations for disease than it does. Did it never catch on in medicine or did it fall out of favor? Gluckman explains that the earliest evolutionary thinkers, around the beginning of the nineteenth century, actually came from a medical background but that the teaching of evolution in medicine became restricted to liberal parts of Europe and prohibited by religious dogma in others. By the end of the century, evolution found itself jostling for intellectual space with new sciences like physiology such that even staunch Darwinists like Darwin's bulldog, Thomas Huxley, believed evolution irrelevant to the problems doctors had to face. Medicine ever since has seen the flourishing of physiology, histology, many other *-ologies*, and biochemistry, which can all be used and taught without any recourse to evolution at all. Evolution comprehensively lost out.

Part of the problem is that many medical professionals are, and always have been, actively hostile to the very idea of evolution. The philosopher Michael Ruse quips that if you want to find the creationists on any university campus, you should head directly to the medical and veterinary schools! Or they find so-called insights into the human condition gleaned from evolution irrelevant to them in daily practice, when constantly faced with seriously ill or dying patients requiring immediate medication or surgery. They live in the proximate world of human suffering, not the ultimate world of evolutionary mechanisms.

Another problem is that a great deal of evolutionary theory and language, as applied to human biology and behavior, seems to grate with our deeply felt understanding of the appropriate moral, ethical, and emotional stances that define humanity. I remember a particularly painful encounter with a woman at a party, while trying to explain studies that claimed to have shown that mothers feed sons richer breast milk than daughters when times are good, and vice versa when times are bad. The idea is that there is an evolutionary mechanism that restricts parental investment in sons in deprived conditions where they may grow up at the bottom of the social pile and be unattractive as mates, limiting the potential for grandchildren. My listener snorted angrily and stomped off in search of more agreeable conversation, tossing out "Why don't you get your facts and your head straight!" over her shoulder as she went, finding the accusation that any decent woman would deliberately withhold breast milk from a suckling child reprehensible and sexist. She had failed to make the distinction between a conscious decision taken against the interests of a child, as in the very unsettling practice of female infanticide in rural India and other places, and an unconscious evolved physiological mechanism to improve the chances of survival of parental genes in either sons or daughters. This reacts to environmental conditions by appropriately channeling infant nutrition from its mother's breasts in a way that requires no conscious thought whatsoever.

Evolution-speak is frequently so dispassionate as to sound positively offensive. It is similarly unsettling to be confronted with evidence that sex ratios can be unconsciously skewed depending on environmental conditions; that nighttime sleep disruption and suckling demands by babies may be an evolved mechanism to prevent their mothers from ovulating and becoming pregnant again, thereby decreasing the likelihood of sibling competition; or that the rosy picture of a loving couple making babies and a woman incubating a fetus and suckling her child is actually an evolutionary battleground of competing male, female, and fetal genetic interests.

For whatever reason, it has taken a century to put evolution back on the map. The 1994 publication of *Why We Get Sick: The New Science of Darwinian Medicine* by Randolph Nesse and George Williams lit the fuse and was followed by important contributions from Peter Gluckman, Wenda Trevathan, Stephen Stearns, Paul Ewald, and many others. (For a full list of who's who in evolutionary medicine, I encourage you to search the growing list of authors and contributors on the *Evolution and Medicine Review* website.) These authors deal extensively and specifically with evolutionary medicine concepts like trade-off and mismatch; however, I want to take a slightly different path that is inspired by a very important point made by one of the fathers of evolutionary medicine, Randolph (Randy) Nesse.

Nesse has frequently claimed that the value of evolution to medicine is that it *may* lead directly to changes in medical practice or indeed to new therapies. However, more fundamentally, its value lies in explaining *why things are as they are*. In that sense, evolution is to medicine as physics is to engineering. Indeed, Nesse's most famous aphorism is: "Medicine without evolution is like engineering without physics." It would be impossible to imagine building the Rosetta spacecraft, sending it 300 million miles to rendezvous with Comet 67P, and successfully deploying the Philae probe, chock-full with sampling instruments, without physics and specifically Newtonian mechanics and a thorough knowledge of the electromagnetic spectrum. It proves similarly impossible to get to the root of the very peculiar human immune system and design really effective cures for allergies and autoimmune diseases without a fresh understanding of how the immune system evolved and for what reasons. Thus, Nesse argues that evolutionary biology should be the foundation and cornerstone for medicine as it should be for all biology. I have known Randy Nesse for well over a quarter of a century and have long admired his persistence, energy, and clarity of argument in trying to drive evolution back into the crowded curriculum of medical school—and recent signs are that the tide is turning in his favor. This book is an attempt to put yet more flesh on the bones of Nesse's idea that evolution is the "physics" of medicine. I am not offering a comprehensive guide to how evolution, across the board, can resolve medical complications, nor am I offering a "heal yourself" manual based on evolution. But I do try to describe the deep evolutionary background to some human medical conditions and to explain why they exist in the first place—why things are how they are. I hope it will leave readers with a new respect for evolution as the prime mover for the structure and function of human bodies, even if it does, on occasion, cause them to break down and drives us into the ER!

In the following chapters, I have tried to supply a satisfying level of deep understanding of the evolutionary factors lying at the heart of a number of disease processes; to dispel a few myths, such as in my discussion of the relationship between our upright stance and degeneration of the spine, feet, and joints; and to show the many ways that coming to an understanding of illness from an evolutionary point of view is already leading to exciting new ideas for medical interventions for blindness, heart disease, autoimmunity, diseases of reproduction, cancer, and Alzheimer's disease. How can heart disease, cancer, or dementia ever be seen as an evolutionary adaptation, you might ask. Well, of course, they can't. But, as I hope to show, the value of thinking in evolutionary terms is that it acts like a forensic tool to ask fundamental questions that allow us to reframe illness in a different way that might then lead to new answers.

For instance, when we look at an X-ray of clogged coronary arteries, it is difficult to escape the thought that these narrow-bore vessels, so prone to blockage, represent an evolutionary design gaffe. We forget that the heart is one of the most powerful, dense masses of muscle in the entire human body and, correspondingly, has its own huge appetite for oxygen and nutrients. Yet, ironically, the denser the heart muscle becomes, the more impenetrable it is to blood supply. We will see that it helps to come to terms with heart disease when we understand why coronary arteries became evolution's answer to driving oxygen-rich blood into increasingly powerful, dense heart muscle in active vertebrates like ourselves. The design brief to reconcile walking upright with giving birth to ever-larger babies has similarly required an inspired compromise in the design of the female human spine and pelvis.

The world of our ancestors was a much dirtier place than it is now. Evolution took the expedient route, since microorganisms in prehistory could not be eradicated, of allowing humans to live with them rather than continually fight them. The great collateral cost of self-inflicted damage to our tissues caused by permanently raging immune systems was avoided by handing over the regulation of our immune systems to the microbes inside us, so that we ended up tolerating them. Evolution could not foresee a world where public hygiene, antibiotics, and chemicals that kill 99.9 percent of all household germs has so depleted this microbial population inside all of us that our immune systems no longer mature properly or are properly regulated, giving rise to dramatic increases in allergy and autoimmunity. This evolutionary mismatch has led to new twenty-first-century epidemics.

There can be no greater argument for an evolutionary approach to medicine than the current waves of antibiotic resistance. Evolution-minded biologists have been warning that this would happen for decades based on the simple observation that bacteria can reproduce in hours or minutes, compared to decades for us humans, and so can evolve at breakneck speed. But we have been deaf to their arguments and allowed decades of fertile and sophisticated antibiotic discovery to come to nothing by willfully overprescribing antibiotics to humans, and making things worse by feeding antibiotics by the hundredweight to animals as growth supplements. We are now in grave danger of being left at the mercy of many multiply resistant and highly pathogenic microorganisms, and health leaders predict a temporary return to hospital wards of the 1950s, complete with widely spaced beds, legions of carbolic-wielding nurses, and wide-open windows wafting in fresh air, while governments try to tempt reluctant pharmaceutical companies with generous tax breaks to reenter the battle against the bugs. Today many oncologists fail to learn the same lesson by understanding that cancer cells behave very much like bacteria and are therefore similarly capable of dramatic and very rapid evolution against the chemotherapy that doctors throw at them. Although survival rates for many forms of cancer have gradually improved, evolved resistance inside tumors is holding back success in many cases.

On the reproduction front, it is difficult to explain away the very low fecundity of humans, compared to other animal species, and the very high rate of spontaneous abortion and diseases of pregnancy like preeclampsia, without a framework from evolutionary theory that takes into account the competing interests of maternal and paternal genes and the evolutionary safeguarding of a mother's investment in each of her offspring.

There is, however, one aspect of evolutionists' description of the human body and its propensity for disease that has always troubled me and impels me to deal with it because I think it serves as a distraction to the evolutionary account of human evolution (and evolution in general). It has come about because of the age-old battle for hearts and minds between Darwinists and the proponents of creationism and intelligent design. Creationists operate from the basic assumption that God has made human beings in his likeness. Darwinists counter this creationist belief by arguing that the human body is riddled with design flaws that no divine engineer-in-chief could possibly be happy with. These multiple imperfections, they say, are the hallmarks of evolution, not divine intervention. This evolutionist/creationist battle can be neatly summed up by my version of a familiar, hoary old joke:

It is the annual meeting of the American Medical Association. The venue is Chattanooga, Tennessee, deep in the Bible Belt, and one of the most Bible-minded cities in America. A group of doctors are relaxing in a lounge, over a few drinks, and the conversation turns to the extraordinary design of the human body. As some members of the party are deeply religious, while a few are secular, they begin to argue over who was the architect of the human body—God or evolution.

"There can be no finer advertisement for the hand of God in human design," pipes up the orthopedic surgeon, "than the human knee. It's the most complex joint in the whole human body. The three long bones of the leg—the tibia, femur, and fibula—all coming together at carefully crafted articulating surfaces, protected by the kneecap, and held together and extended and retracted by a plethora of tendons and ligaments, with cartilage shock absorbers and fluid-filled sacs to make it move smoothly. Miraculous!"

"That's all well and good," counters the neuroscientist. "But you have to look at the human brain in all its bewildering complexity to fully appreciate

God's handiwork. Just think about it—86 billion neurons all sending nerve impulses to each other at up to 260 miles per hour, through a vast network of 125 trillion synapses! I'm working on the computerized mapping of brain activity, and we reckon we'll need 300 *billion* gigabytes of computer memory just to store one year's worth of measurements!"

"Well, I don't know about that," says the urologist. "Down below the waist, where I do my work, I tend to see things a bit differently. There's some pretty flawed plumbing all over the place. What about the ridiculous long-winded route the vas deferens has to take between the testis and the penis because it has to loop up and over the ureters? And the vise-like grip the encircling prostate gland has on the urethra as it exits the bladder that makes it so difficult for middle-aged men to pee properly? What a botched job! All that betrays the hand of evolution to me. And while we're talking about the urino-genital system, what divine creator would be so idiotic as to lay a sewage pipe straight through the middle of a recreation area? It's a miracle, all right—a miracle of bad engineering!"

The evolutionary literature is strewn with other examples: The pharynx is used both for respiration and ingestion of food, greatly increasing the chance of choking. The human appendix is a vestigial organ, yet its continued presence leaves us susceptible to appendicitis that was a killer before the intervention of modern medicine. We humans suffer from poor sinus drainage because our faces have flattened and we now walk upright. Sinuses that used to drain forward now drain upward. And, in other animals, the example of the route of the recurrent laryngeal nerve is a great favorite among evolutionists. The nerve connects the larynx to the brain, but it has to loop around the aortic arch in the process. Thus, the longer the neck, the longer the nerve has to be. In giraffes it is over twenty feet long. Surely a divine perfectionist would have disconnected and rerouted it?

This is called the argument from poor design, and the problem with it is that it tends to portray evolution as a rather hapless botcher—a jerry-builder—whose witless quick fixes lack the simplicity and elegance of good design and resemble instead the fantastic overelaborate yet dysfunctional complexity of either Rube Goldberg or Heath Robinson—depending upon which side of the Atlantic you are arguing from. Thus Nesse and one of the world's most famous evolutionists, Richard Dawkins, jointly criticize the design of the human eye, whose inverted retina makes light traverse a jumble of cells before intercepting the photoreceptors, and traipses the wiring carrying visual signals across the surface of the retina before plunging them through the blind spot en route to the brain. What a messy solution!

The real tragedy, for me, with all these jerry-rigging metaphors, largely born out of the sort of twisted logic that tends to be employed against creationists, is that, swallowed whole, they do evolution a great disservice. Instead of emphasizing the peculiarity of evolutionary design, they end up giving us a portrait of evolution as an inefficient meddling tinkerer and sometimes neglect to remind us that evolutionary design solutions for our bodies are, in their own ways, exquisite and functional. But they are design solutions arrived at by blind, future-ignorant processes of mutation and selection, and are therefore necessarily idiosyncratic and, to pure-minded engineers, sometimes eccentric. After all, the products of evolution in our bodies cannot simply be ragbags of fanciful and ineffective confections. Had evolution consistently made a hash of it, our species would have perished long ago in numerous evolutionary arms races. In that sense, evolutionary design of the human body is more likely to call to mind the resourceful Angus MacGyver of the television series *MacGyver*, who was always fashioning inspired solutions to life-or-death problems out of everyday items like duct tape and paper clips, than the incompetent and anatomically ignorant quack physician Dr. Nick of *The Simpsons*, who famously, and witlessly, botched a transplant operation to give his patient an arm where a leg ought to have been, and vice versa!

I think it is a pity that evolutionists have boxed themselves into the position where, to convince us that evolution, not God, designed the human body, they have found themselves accentuating the negative. Take the design of the eye, for instance. I argue that had some evolutionists dug a little deeper, they would have discovered very plausible reasons why the evolutionary design of the inverted retina, which superficially looks flawed, could well be a beautiful and powerful adaptation to allow vast amounts of computation of visual signals. I would like to see the argument from poor design thrown in the trash. In the case that concerns us here—evolutionary medicine—it weakens, rather than strengthens, the argument.

Evolutionary design, as illustrated in our bodies, is frequently inspired rather than inept. But millions of years of human evolution have left indelible marks on our bodies, not all of which we perceive today as being positive. We've survived and thrived as a species, but our bodies are littered with all the trade-offs that evolution has made, all evolution's quick-and-dirty fixes, all the "live now, pay later" antagonistic pleiotropies where evolution has invested in mechanisms designed to keep young people alive into reproductive age at the expense of the negative effects on health as we get older, all the unintended consequences of evolutionary change, all the mismatches between our evolved bodies and modern environments, and all the compromises, sometimes of almost Faustian proportions, that evolution has engineered. All these effects we interpret today as sickness and disease. Today we are constantly bombarded by popular keep-fit literature titled "Body by This" or "Body by That"—you can fill in the blanks for yourself: veganism, God, science, Vi, or the personal trainer of your choice. Well, this is your body thanks to evolution by natural selection, or, to put it figuratively—and by deliberately conflating the theory with the theorist—this is your body by Darwin.

And at the heart of all these areas of illness lie countless individuals trapped in the disease process in question. I have incorporated their voices lest we forget that, among all the abstractions of evolutionary theory, real peoples' lives are touched by illness and infirmity, real people suffer, and real people constantly show enormous fortitude in enduring disease or caring for those who do. Many of the people I have interviewed are bravely acting as guinea pigs for pathbreaking therapies, inspired by evolution, for their disease. I want to thank them all for their unstinting help.

ABSENT FRIENDS

HOW THE HYGIENE HYPOTHESIS EXPLAINS ALLERGIES AND AUTOIMMUNE DISEASES

Throughout the 1990s the Johnson family was being ripped apart by the increasingly violent, self-abusive, uncontrollable behavior of their son Lawrence. He was a very disturbed child and would become highly agitated, smash himself in the face, bang his head against walls, try to gouge out his eyes, and bite his arms until he bled. At age two and a half, he was diagnosed with autism, and as he got older, things got worse. If traffic lights failed to change according to his inner timetable as he passed them while walking down the street, he would explode with rage. He could not deal with crowded places like restaurants or movie theaters, and frequently had to be physically restrained from hurting himself. His doctors tried anti-depressive medication, anti-seizure drugs, anti-psychotics, and lithium, among others, to no avail.

His parents were at their wits' end. But because Lawrence's father, Stewart, is an active, coping, problem-solving sort of man, he tried to think his way through his son's illness and became a self-taught scholar on autism. A clue soon emerged. "We noticed that when Lawrence had a fever, all that would go away. And it was true 100 percent of the time. If his temperature went up and he had a fever from a cold, flu, or sinus infection, he would stop hitting himself, he would be calm, he was like a different child. We talked to other parents with autistic kids, and they all said the same thing."

Was it simply the lethargy of feeling poorly that damped down Lawrence's bad behavior? Some scientists have suggested that fever changes neural transmission in the brain, while others have invoked changes in the immune system. No one knew for sure. But everybody who either had to live with Lawrence or care for him said the same thing: "We're very happy when he gets sick, because life is wonderful!" However, whenever his fever subsided, the frightening behavior returned. By 2005, when Lawrence was fifteen, Stewart and his wife decided they could no longer take care of Lawrence on their own. While Lawrence was away at a specialized summer school, they reluctantly applied to have him taken into care for the rest of his life. "Lawrence was going to go and be somebody else's problem because he was killing our whole family."

It was at that precise moment, with Lawrence's dismal future decided, that they received a phone call from summer camp. They feared the worst. "But they said, 'We don't know what's going on, but Lawrence is behaving completely normally. He's fine, he's not freaking out, he's not hitting himself, he's not throwing his food, he's participating in all the activities, he's interacting. . . .'" Stewart drove up to the camp to find his son perfectly calm, engaged with the other children, and pleased to see him. They got into the car for the two-hour drive home, and Lawrence announced that he wanted to go out for dinner. He hadn't been to a restaurant in two or three years. "But now he wants to go to this place that's extremely noisy and crowded, which is a place we'd normally cross the street to avoid. He would normally never wait in line, but we waited forty-five minutes, we got served, he ate, we had a wonderful dinner!"

Stewart drove the family home, his mind whirling. Later that night he was helping Lawrence get undressed for bed when he noticed that he was completely covered in chigger bites from his thighs to his ankles. Chiggers, biting mite larvae, are very common in the summertime where the family lived, and Lawrence had been bitten by the chiggers in the long grass of summer camp. What could be the link between chigger bites and the total remission of Lawrence's autistic symptoms? It drove Stewart back into the literature to discover that chiggers elicit a very powerful immune response as they drill into the skin and release saliva to digest host tissue. They then drop off, leaving a patch of hardened scar tissue that itches for days until the immune reaction subsides. For ten days, while Lawrence's immune system fought the chigger infection instead of him, they had a wonderful time, but as the bites faded and the itching stopped, he returned to his violent and self-destructive behavior. "I immediately said, 'That's it. I've seen enough. There's an answer here. At least some part of my son's symptoms are an aberrant immune response.'"

Stewart knew that his son's physician, autism expert Dr. Eric Hollander of the Albert Einstein College of Medicine in New York, had done research showing that there was nine times the incidence of autoimmune disease in the first-order relatives of autistic children, compared to those of normal children. Lawrence has a peanut allergy; Stewart suffers from myasthenia gravis, an autoimmune disorder that causes fatigue and weakened muscles; and his wife is asthmatic. His family medical history was consistent with research that links autism to autoimmunity and allergy. Back in 1971, researchers at Johns Hopkins University reported on a family study where the youngest son in the family had a multiple diagnosis of autism, Addison's disease (an autoimmune condition affecting the adrenal glands), and candidiasis (infection with the opportunistic yeast, *Candida albicans*). The next older brother had hypoparathyroidism, which can have autoimmune origins, Addison's disease, candidiasis, and type 1 diabetes. The next older brother had hypothyroidism, Addison's disease, candidiasis, and alopecia totalis—an autoimmune condition that causes the loss of all head hair. The oldest son, and firstborn, was symptom-free, like his parents.

In 2003 Thayne Sweeten, from Indiana University School of Medicine, reported that the rate of autoimmune disorders in the families of children with autism was even higher than in the relatives of children with autoimmune diseases. Disorders included hypothyroidism, Hashimoto's thyroiditis (where the thyroid gland is attacked by autoantibodies and immune cells), and rheumatic fever. Sweeten said that the finding of increased autoimmune illness in grandmothers, uncles, mothers, and brothers of autistic children "suggests a possible mother-to-son transmission of susceptibility to autoimmune diseases in the families of autistics," and speculated that autoimmunity or chronic immune system activation could account for some of the biochemical anomalies seen in autistic individuals, including high uric acid levels and iron-deficiency anemia, which are also seen in autoimmune disorders. Research conducted on Danish children between 1993 and 2004 by Dr. Hjördis Atladóttir agreed with Sweeten by finding higher rates of autism in children born to mothers who suffered from celiac disease (where you cannot tolerate gluten). The study also linked autism to a family history of type 1 diabetes and to children whose mothers suffered from rheumatoid arthritis.

Chiggers, remission of autism, and autoimmunity all began to come together in Stewart Johnson's forensic mind. If immune disorder—a hyperactive immune system—was causing his son's autism, he needed something to damp it down. His research led him to the work of Joel Weinstock, David Elliott, and colleagues, then at the University of Iowa. Weinstock's group had reported success in medical trials where they had treated a small number of patients with Crohn's disease, an autoimmune inflammatory bowel disorder, with the eggs of an intestinal parasite—the pig whipworm. They had treated one group of twenty-nine patients with 2,500 live whipworm (*Trichuris suis*) eggs, delivered via a tube into their stomach, every three weeks for twenty-four weeks. By the end of the treatment period, 79 percent of the patients had responded dramatically; the whipworm eggs had driven their Crohn's disease into remission. "I was impressed," says Stewart. "These were real scientists doing real work and getting results; this was not fringe stuff. It certainly looked good for Crohn's, so I thought, 'Maybe I'm right with this.' So I wrote all this up and did a sort of mini research paper with references, and I gave it to Eric Hollander."

Hollander was intrigued. "Stewart is a very smart guy and he'd done a lot of intense research and he just pulled out the literature and we talked it over. It seemed like a plausible hypothesis and a reasonable thing to try." Hollander obtained the necessary clearance to administer the treatment and helped Stewart with the task of importing the whipworm eggs from Germany. They began with a low dose, for fear of side effects. Stewart took the eggs as well—he wasn't about to foist a seemingly bizarre treatment on his son that he was not prepared to share. The initial results were extremely disappointing. They only saw the "good" Lawrence for four noncontiguous days during the whole twenty-four weeks of treatment. Stewart rang the manufacturers and they told him that the results were actually predictive of someone who ultimately does respond and will respond better at a higher dose. So they went to the same level that Weinstock had set for his Crohn's patients—2,500 eggs per treatment. Within eight days Lawrence's symptoms completely evaporated, and they have stayed away ever since. The old Lawrence only briefly returned four times when they experimented by taking him off the eggs for a few days. So far, as long as he keeps taking the eggs, his autism symptoms appear to be kept at bay.

Stewart Johnson had discovered the hygiene hypothesis, which links the bacteria, fungi, and helminths (parasitic worms) in our guts, on our skin, and in our airways and vaginas, and a host of autoimmune and allergic diseases. There is mounting evidence that the composition of all these organisms, living on us and inside us—collectively known as our microbiota—can offer protection against a formidable list of autoimmune diseases, including the inflammatory bowel diseases Crohn's disease and ulcerative colitis, type 1 diabetes, rheumatoid arthritis, multiple sclerosis, and, as we have seen, mental health. Some studies suggest that they can also protect against a comparable list of common atopic or allergic illnesses like eczema; food, pollen, and pet allergies; hay fever, rhinitis, and asthma. Nevertheless, it cannot be stressed strongly enough that autism is a complex, multifactorial illness and that the therapeutic application of all the science associated with the hygiene hypothesis, for a variety of autoimmune and allergic diseases, is still very much in its infancy and largely unproven—Lawrence Johnson's treatment, for instance, is a one-off experiment, not the result of tried-and-tested medicine. But much of the research is compelling, and if it can translate into medical therapy, it could represent, within a few years, nothing short of a revolution in medicine.

Dramatic improvements in hygiene, sanitation, and water quality from late Victorian times to the present day, allied to extensive use of antibiotics and population-wide vaccination, have raised the quality of life and life expectancy throughout the developed world. But, perplexingly, post-industrial society—while largely eradicating epidemics of polio, whooping cough, dysentery, measles, and many other potentially lethal or debilitating infections—has fallen prey to new, major growing epidemics of autoimmune and allergic disease. Take bowel disease for example. According to Weinstock's research, prior to the twentieth century, inflammatory bowel disease (IBD) was largely unknown. Between 1884 and 1909, hospitals in London were averaging two cases of ulcerative colitis per year at most, and Crohn's disease only became recognized in 1932. But during the second half of the twentieth century, IBD gained in scope and prevalence. Currently, IBD in the United States affects between 1 and 1.7 million people. The current estimate is that 2.2 million people in Western Europe and the United Kingdom have IBD. Although once thought to have stabilized, the incidence of Crohn's disease continues to gradually rise in England, France, and Sweden. IBD is less prevalent in Eastern Europe, Asia, Africa, and South America. However, as countries in these regions develop socioeconomically, IBD increases. Moreover, when people move from a country with a low prevalence to a country with high prevalence of IBD, their children acquire a higher risk of developing IBD.

Although type 1 diabetes has been around for centuries, it also is on the rise—and too fast for genetic change to be implicated. This guilt-by-association link between affluent, hygienic, Westernized countries and the rise of autoimmunity also extends to multiple sclerosis (MS), which has a low incidence in tropical regions that increases as you move north of the equator. In the United States, prevalence is twice as high north of parallel 37 than below. Infectious agents, genetics, and vitamin D levels have all been implicated, but, intriguingly, adult immigrants leaving Europe for South Africa have a threefold higher risk of developing multiple sclerosis than those migrating at age fifteen or less, suggesting a protective environmental effect in the adoptive country, operating only on the young. The opposite trend is seen among the children of immigrants to the United Kingdom from India, Africa, and the Caribbean (all regions of low prevalence), where the risk of developing MS is higher than among their parents but similar to children born in the UK. Jorge Correale, a neurologist from Buenos Aires, points out that MS is rising steadily in all developed nations. In Germany, the incidence of MS doubled between 1969 and 1986, he says, while MS has increased twenty-nine-fold in Mexico since 1970, in line with steadily improving living standards. Correale also reports a fascinating inverse relationship between MS and the distribution of a very common intestinal parasitic whipworm, *Trichuris trichiura*, which used to be extremely common in the southern United States and still is common throughout the developing world. MS prevalence, he explains, falls steeply once a critical threshold of 10 percent of the population infected is reached. In a similar way, common atopic diseases like eczema and asthma are relatively rare in the developing world, whereas levels of helminth infection are relatively high.

Weinstock recalls how the "penny dropped" for him while ruminating in his seat on an aircraft waiting interminably for takeoff from Chicago's O'Hare Airport. He was thinking of cause and effect—I do something and then something happens. He suddenly realized that the answer to the conundrum of rising levels of bowel disorder and other autoimmune diseases lay in "what doesn't happen but used to." In other words, it wasn't about what new aspects of our environment could be contributing to autoimmunity but what had been taken away from our modern environment that might leave us open to it. "In the historical environment we had filthy streets—horse dung was a major feature—and many people walked barefoot or were poorly shod. But now we have built roads and sidewalks and we wear proper shoes, and so our ability to transmit these organisms from one another went down lower and lower. Then we cleaned up the food supply, the water . . . everything got clean. As a result worms disappeared. And when you look at the incidence of deworming from population to population and the rise of immune-mediated diseases, you find that they are inversely related. That's a negative correlation, it doesn't prove that worms are effective—but it is a smoking gun."

Modern sanitation has proved disastrous to most helminths, says Weinstock. Indoor plumbing and modern sewage treatment spirit away their eggs before they can spread infection, as do frequent baths and laundered clothing. Cleaning fluids disinfect utensils and domestic surfaces, blocking their transmission. Sidewalks and shoes obstruct the common hookworms *Necator americanus*, *Ancylostoma duodenale*, and *Strongyloides stercoralis*, while modern food processing kills *Diphyllobothrium*, *Taenia*, and *Trichinella* larva. These changes have all but eradicated helminths from industrialized countries. Until the 1960s, trichinosis was endemic in the northeastern and western United States through the eating of contaminated pork. There are now less than twenty-five cases a year. Removal of these parasites has undoubtedly reduced a huge amount of morbidity in the population, but the baby has been sluiced away with the bathwater—the protection afforded by these organisms has gone with it. A classic example of this two-edged sword effect comes today from East Africa. Educational learning specialists looking at achievement in Kenyan schools had assumed that lack of textbooks and flip charts were the major factors holding pupils back but discovered, to their surprise, that helminth gut parasites were much more important. Huge deworming programs have since almost eradicated bilharzia and hookworm, and exam results have consequently risen dramatically. However, the unwanted by-product of deworming has been a dramatic rise in eczema and other allergies among Kenyan and Ugandan children. In tropical Africa the irritation of these skin conditions often goes untreated, and constant rubbing and scratching of the sores leaves children open to infection and septicemia.

Correale treats multiple sclerosis patients in Argentina. Out of a group of twenty-four patients in whom he was monitoring the progression of the disease, he identified twelve patients who were carrying mild loads of intestinal parasites. He followed them for a little over four years, regularly checking their immunological function and performing MRI scans to identify lesions in their brains and spinal cords. The infected individuals had significantly fewer relapses, fewer lesions, and better measurements on all disability scores. He extended his observation to seven years, but after five years, four of the patients had to be given anti-helminthic treatment—their worm infections were causing gut pain and diarrhea. As soon as the helminths were swept out of their systems, all the MS indicators rose and they were soon undistinguishable from uninfected patients.

Erika von Mutius, an expert on allergies from Munich University, had a unique opportunity in the unification of East and West Germany to test her theory that high levels of air pollution and crowded, poor living conditions would lead to higher levels of asthma, hay fever, and other atopic illnesses. She expected that children from wealthier West Germany—with its cleaner environment, high levels of sanitation, and lower levels of polluting heavy industry—would show less atopic illness than children in East Germany. To her surprise, the opposite was true. East German children—who characteristically lived in cramped domestic conditions surrounded by lots of siblings, pets, and other animals and who spent longer hours in day care—showed far less allergy and asthma than their West German counterparts. Her revised conclusions are that early exposure to a variety of childhood microbial infections from brothers and sisters, other children, and animals somehow conditions the immune system such that it becomes more tolerant of potential allergens later in life.

She has since gone on to compare urban and rural communities all over Europe to show that children who grow up on traditional farms—where they come into contact from birth onward with livestock and their fodder, and they drink unpasteurized cow's milk—are protected from asthma, hay fever, and other allergic sensitizations. Von Mutius explains that throughout Switzerland, Austria, and Germany, where traditionally farming has been the main source of subsistence, most farms are involved in dairy production but may also keep other animals such as horses, pigs, and poultry. In addition, some farmers raise sheep and goats. Most farmers also grow grass, corn, and grain for fodder, and many farmhouses place fodder, people, and animals close together under one roof. Furthermore, women work in stables and barns before, during, and after pregnancy, and children a few days old are taken into stables, where mothers can look after them while working. Von Mutius stresses that several things seem of paramount importance for inducing tolerance to allergens. Exposure to microbes very early in life, even in utero, is vital, as is the diversity of animal species with which children come into contact and directly relates to the richness, in number and species diversity, of microbes.

Of all the autoimmune diseases, type 1 (or early-onset) diabetes is rapidly becoming the main scourge of life in the modern, hygienic Western world. Rates are set to double over the next decade among European children under the age of five. But the hot spot is Finland, with the highest rates of type 1 diabetes in the world. In an effort to find out why, Mikael Knip and colleagues, from the University of Helsinki, have been conducting an extraordinary population survey designed to disentangle genetics and environment in the causation of this life-threatening disease, where the body attacks the insulin-producing beta cells of the pancreas, causing high blood sugar. Although insulin treatment has saved lives and can stabilize the condition, many people suffer in the long-term from blindness and kidney damage.

Karelia, the old home of the Karelian people, is a large northern European landmass that used to belong to Finland but was partly ceded to Russia during World War II. As a result, the country has been partitioned into Russian and Finnish Karelia ever since. Although Russian and Finnish Karelians have the same genetic makeup, including the same susceptibilities to diabetes, the differences in their socioeconomic status and health could not be more stark. According to Knip, one of the steepest standard-of-living gradients in the world exists at the border between Russian and Finnish Karelia, with the latter having eight times the gross national product of the former. That is even greater than the difference between Mexico and the United States. Yet the incidence of type 1 diabetes, and a host of other autoimmune diseases, is far higher on the Finnish side. Finnish Karelians have six times the incidence of diabetes, five times the frequency of celiac disease, six times more thyroid autoimmunity, and much higher allergy levels than Russian Karelians.

Knip managed to get cooperation from the Russian medical authorities and collected medical data, stool samples, blood samples, and swabs from skin and nasal passages from thousands of children on both sides of the border. They found that Russian Karelians experienced much higher loads of microbial infection by the time they were twelve years old, and they tended to have more diverse colonies of microbes in the gut, where bacterial species known to play an active role in protecting and maintaining the gut lining were more common. They also noted biochemical evidence of much better regulated immune systems. Vitamin D deficiency has often been cited as a causative agent for type 1 diabetes, yet the researchers found that vitamin D levels were generally lower on the Russian and Estonian sides of the border than in Finland, with its much higher rates of the disease. To put it bluntly, Russian Karelians are poorer and dirtier than their counterparts on the Finnish side, but in terms of immune-related diseases, they are much healthier.

Does early exposure to a variety of bacteria, fungi, and helminths (which, traditionally, we would have inhaled, been infected by, or ingested from birth) work the same way as a childhood vaccination—such as the triple vaccine against measles, mumps, and scarlet fever—in stimulating immunity? The original form of the hygiene hypothesis suggested that it did. The hypothesis began in the nineteenth century in the context of allergies. In 1873 Charles Harrison Blackley had noted that hay fever was associated with exposure to pollen but that farmers rarely experienced the condition. Later, in the 1980s, David Strachan, from St. George's Hospital in London, observed that having many older siblings correlated with a diminished risk of hay fever. The assumption was that this "grubby brother syndrome" of postnatal infections, rife in large families, protected against allergies. Strachan's theory, therefore, suggested that the immune system is primed by early exposure to develop acquired immunity to these diseases, just as occurs with childhood vaccination, and that our almost pathological obsession with hygiene in the West has removed these important stimuli. However, over the last ten years, a number of clues have arisen that suggest there is something much more profound going on.

The first clue lies in the huge length of evolutionary time that we humans have been exposed to certain bacteria, fungi, and helminths, compared to more modern pathogens like cholera and measles. George Armelagos, of Emory University, says that between the Paleolithic period, some 2.5 million years ago, up to about 10,000 years ago, our human ancestors would have been frequently in contact with saprophytic mycobacteria that abound in soil and decomposing vegetation. Paleolithic diets would likely have contained a billion times more of these non-pathogenic bacteria, like lactobacilli, than do diets today because of the unprocessed food our ancestors ate and because they stored food in the ground. They would also have been exposed to chronic infection by a variety of helminth worms. Molecular analysis of tapeworms, explains Armelagos, shows that they were ubiquitous parasites in human guts 160,000 years ago, before the out-of-Africa human exodus. A severe hookworm infection would cause poor health in any human, but it would be unlikely to kill its host. Once established, most helminthic infections would have been almost impossible to dislodge before the advent of modern drug treatment, and a hyperaggressive immune response against them would eventually do far more damage to the human host than to the worms—they had to be tolerated.

Only with the establishment of the first cities over six thousand years ago did humans transition to more crowded settlements, and a new raft of serious epidemic diseases—cholera, typhus, measles, mumps, smallpox, and many others—arose. These more modern diseases have not been around long enough to cause as much evolutionary change in us as more ancient infections. Helminths, fungi, mycobacteria, and commensal bacterial species have been part of our "disease-scape," as Armelagos coins it, for eons. We, and they, have had time to evolve together—we have coevolved. Graham Rook, from University College London and a doyen of the field, not surprisingly calls these organisms "old friends" and has renamed the hygiene hypothesis, having given it this deep-time coevolutionary dimension, the "Old Friends Hypothesis."

The second clue to this extraordinary coevolution lies in the fact that we need early exposure to these "old friend" microorganisms not merely to activate our immune systems but to establish, build, and mature our immune systems in the first place. There is no finer example of how we humans have coevolved with the bugs inside us than our interaction with bacteria during childbirth and the first crucial months of life.

During pregnancy there are important changes in the spectrum of bacterial species that populate the vagina. Both numbers of species and total numbers of bacteria decrease, but, within that, several species enrich their presence. Most of them are *Lactobacillus*—the genus of bacteria that commonly make up our probiotic yogurts, specifically *L. crispatus*, *L. jensenii*, *L. iners*, and *L. johnsonii*. They keep the mucosal lining of the vagina slightly acid, which protects it against pathogens, but they are also important members of a stable gut microbiota; when they are swept into a baby's mouth and involuntarily ingested as it pushes its way into the world, they will rapidly populate the infant gut, protecting it against pathogenic bacteria like *Enterococcus*. As the baby exits the vagina, it also becomes "infected" by bacteria present in traces of feces from its mother's anus. Particularly common species are a number of facultative anaerobes (bacteria that can operate in the presence of oxygen but also switch to fermentation in anaerobic conditions). They are able to establish themselves in the infant gut very early on because it contains oxygen, but they may also rapidly drive the gut toward a state where obligate anaerobes, including one of the principle "friendly" bacteria, *Bifidobacterium*, can thrive and take over.

The baby is born with a gut that is almost completely sterile and must be populated immediately with bacteria. If it is breast-fed, it starts to receive one of the most extraordinary products in the natural world. Human breast milk contains a complex array of fats and sugars—fast food—but it also contains immunoglobulin A, an antibody that protects the lining of the human gut and prevents pathogens from attacking and perforating it. It has also been estimated that a breast-fed infant receives over 100 million immune cells every day, including macrophages, neutrophils, and lymphocytes, together with a host of cytokines, chemokines, and colony-stimulating factors—molecules that signal between cells of the immune system and promote their growth. Over seven hundred species of bacteria have been found in human breast milk, many of them—like *Lactococcus*, *Leuconostoc*, and *Lactobacillus*—are capable of digesting milk sugars. Bifidobacteria, members of one of the most potent probiotic genera, are also predominant.

Some of the main solid components of breast milk are very complex, long-chain sugars called oligosaccharides. They weigh in at approximately ten grams in every liter, and human breast milk contains between ten and a hundred times more oligosaccharides than any other mammal milk tested. Yet the baby is totally incapable of digesting these molecules—it simply doesn't have the enzymes to do the job. Why human breast milk contains such high amounts of indigestible material has perplexed scientists for years, but it is now known that they are never destined for the baby at all. They are specifically produced to nourish the bifidobacteria that accompany them in breast milk. *B. longum*, for instance, has seven hundred unique genes that code for enzymes involved in breaking down oligosaccharides. These friendly bugs are parachuted into the baby's gut along with their own exclusive packed lunch. It gives bifidobacteria a head start in the rough-and-tumble of bacterial competition in the infant gut. Babies' guts rapidly digest and absorb all the simpler sugars so that practically the *only* sugar molecules left undigested by the time food reaches the large intestine will be the oligosaccharides. This means that all the bacterial species there are forced to compete for this sole source of carbon. It is the exclusive enzymatic tool kit of bifidobacteria that gives them the competitive edge. Breast-fed babies, in Darwinian terms, are ecological niches for bifidobacteria. Babies, in turn, have evolved to take full advantage of these friendly bugs. One Finnish study, for instance, has shown that bifidobacteria represent 90 percent of the gut microbiota in three- to five-day-old infants.

Probiotic bacteria like this have crucial functions. The immature, sterile gut of a newborn baby is at the mercy of aggressive pathogens, and its naive immune system is not yet developed and programmed to repel invaders. Probiotic species can act as decoy receptors to bamboozle pathogens and can prevent microbial pathogens from sticking to the gut wall. They seem to protect against necrotizing enterocolitis, for instance. They also help the establishment of a rich biofilm of probiotic-laden mucus that protects the gut, and they direct the development of a well-regulated immune system. The gut has its own immune system, distributed throughout the gut wall, known as the gut-associated lymphoid tissue, and probiotic species have been shown to be vital for its proper development. A number of experiments illustrate how potent this effect is. When oligosaccharides are experimentally added to infant-food preparations in controlled trials, they lower circulating levels of immunoglobulin E (IgE), which is an important marker for allergies. They also reduce atopic dermatitis, diarrhea, and upper-respiratory-tract infections. This baby/oligosaccharide-laden breast-milk/probiotic microbes system is one of the most elegant examples of co-evolution that science has ever uncovered and has been essential for the survival of babies across evolutionary swathes of time.

At the risk of fashioning yet another stick to beat modern mothers over the head with, a number of studies point to the downside of both Cesarean section and bottle-feeding. Babies born via elective Cesarean section are more likely to be initially colonized by bacteria we commonly associate with the skin, and they experience lower initial colonization with "friendly" bacteria like *Bifidobacterium*. They take over five months to establish a stable, healthy microbiota. Bottle-fed babies show higher counts of clostridia, enterobacteria, *Enterococcus*, and *Bacteroides*, which can all be opportunistic pathogens. Dr. Christine Cole Johnson, of the Henry Ford Hospital in Detroit, followed over one thousand babies out to two years old. C-section babies were five times more likely to develop allergies than babies who had been delivered normally. Research elsewhere adds celiac disease, obesity, type I diabetes, and even autism to the risk list. One obvious solution that might counter these early disadvantages would be to deliberately "infect" C-section babies with probiotics at birth. This is exactly what Maria Dominguez-Bello has done, together with colleagues from the United States and Puerto Rico. They have incubated gauze swabs for one hour in the vaginas of women electing for C-section and then painted the babies—first in the mouth, then on the face, and finally all over the rest of the body—with the swabs as soon as they are surgically removed from the womb. They have shown that the babies acquire their mothers' vaginal bacterial populations and show high bacterial species diversity in the gut after birth, which decreases somewhat once breastfeeding begins until it resembles the pattern of microbes present in mother's milk.

It also seems that vertical transmission from mother to baby via breast milk can pass on either good or bad traits directly from one generation to another. There is, for instance, a link between maternal obesity and microbial species diversity in breast milk. Obese mothers produce species-impoverished milk compared to leaner mothers. Their milk is poorer in beneficial probiotic species, and there are higher counts of potentially pathogenic bacteria like *Staphylococcus* and *Streptococcus*. There is evidence that if you receive the gut microbiota that is associated with obesity, you are at higher risk of becoming obese yourself and of developing insulin resistance. Allergic mothers can pass their aberrant immune settings to their babies. The breast milk of allergic mothers contains lower counts of probiotic bifidobacteria. Although several months down the line, breast-fed and bottle-fed babies will stabilize around a similar mix of microbial species in their gut microbiota, it is the pattern of early colonization, in the first few days, that seems vital for the proper direction of the immune system.

How do "friendly" bacteria get into breast milk in the first place? According to Christophe Chassard, from Switzerland, there is a spectrum of bacterial species that is common to the mother's gut, her breast milk, and the baby's gut. It appears that bacteria are actively translocated from gut to breast by crossing the gut wall and entering mesenteric lymph nodes, from which they are transported through the lymphatic system to the mammary glands. This completes the cycle of vertical transmission from mother to baby. However, it does suggest that the baby is largely dependent upon the gut health of its mother. If she has a healthy gut microbiota, the baby will quickly benefit; but if she has a species-poor depleted microbiota, her baby will be similarly compromised.

It took a restaurant owner in London's Covent Garden to realize that human breast milk could be a big seller. He began marketing a brand of ice cream called "Baby Gaga," churned with generous donations from a prodigious nursing mother, Victoria Hiley, together with Madagascan vanilla pods and lemon zest. You could have it served, reported *BBC News*, with the optional extra of a rusk and a shot of Calpol (an infant pain reliever) or Bonjela teething gel. "Some people will hear about it and go 'yuck,' but actually it's pure organic, free-range, and totally natural," said Hiley, while the restaurateur, Matt O'Conner, added: "No one's done anything interesting with ice cream in the last hundred years!" Of course it couldn't last. Westminster Council's food inspectors descended and demanded the refrigerators be cleared, as they couldn't be sure it was fit for human consumption!

Within a week or so after birth, the infant gut, originally sterile, has become colonized by up to 90 trillion microbes. The number of microbes in our guts eventually exceeds the total number of cells in our bodies by a factor of ten; our gut microbiota weighs considerably more than either our brain or liver; and the total number of microbial genes exceeds the number of genes in the human genome by a factor of a hundred. These microbes are not transit tourists, passing through, but long-term residents. Although it has long been recognized that much of this microbiota is benign, it was conventionally assumed that we, and the bugs inside us, simply dine at the same table. We passively allow the bugs to take a proportion of the nutrients flowing through our gut and give them somewhere warm and oxygen-free to live, while they feed us scraps from their digestion, like vitamins B, H, and K, which we cannot manufacture ourselves, and break down sugars like butyrate, which help our metabolism. But it has become clear that our relationship with our "old friends" goes far beyond such symbiosis. We have evolved such a close mutual interdependence with our microbiota that it no longer makes sense to distinguish between the two genomes. Scientists now refer to the existence of a meta-genome to represent the combined genomes of human and microbiota, a superorganism in which we humans are the junior partner and without which we could no longer exist. Scientists are asking two linked and fundamental questions: First, how can our bodies tell the difference between "old friends" (commensal bacteria, fungi, and intestinal worms) and dangerous pathogens, so that they tolerate the former while attacking the latter? Second, what happens to human health when these "old friends" are absent or depleted? This is what is allowing them to close in on the full story of the evolution of this mutual codependency, the development of the human immune system, and a pharmacology of the near future geared toward removing the vast epidemics of allergic and autoimmune disease currently ravaging Westernized societies.

To understand how the "old friends" manipulate our immune systems in order to masquerade as "self," we need a few basic facts about what the immune system is and how it is organized. We humans have two immune systems: the innate immune system, common to the whole animal kingdom—both invertebrates and vertebrates—and an adaptive immune system that only exists in vertebrates. The innate immune system reacts to pathogens in a totally nonspecific way—it cannot offer long-lasting, protective immunity because it has no memory of past insults. It leaps into action whenever a pathogen is sensed, by producing an inflammatory reaction at the site of injury or infection. This serves to cordon off the infected area, dilates the surrounding blood vessels, and recruits a number of immune cells to the site to fight the infection. The inflammation is caused by cytokines—molecules that pass signals between immune cells—together with histamines and prostaglandins. The most important "pro-inflammatory" cytokines are tumor necrosis factor-alpha (TNF-α), interferon-gamma (interferon-γ), and interleukins 1, 6, 7, and 17. The innate immune system also includes the complement system of blood plasma proteins that helps or complements other immune factors by attacking and disrupting pathogens, labeling them so that they can be targeted by other cells, and recruiting more inflammatory factors to the battlefield.

The main cells of the innate immune system are collectively known as leukocytes (white blood cells), but there are many different types. Mast cells are common in all mucosal surfaces, like the lining of the gut and lungs, and release histamine, cytokines, and chemokines—a type of cytokine that acts as a signpost to other immune cells and directs them to the site of action. Most important are phagocytes, a group of active scavengers of pathogens that includes macrophages (literally "big eaters"); neutrophils, which come laden with killer chemicals like hydrogen peroxide, free radicals, and hypochlorite—nature's bleach; and dendritic cells, particularly common in the gut wall, whose main job is to engulf the foreign proteins that make up the coats of pathogenic bacteria and viruses and then "re-present" them on their cell surface in a form that can be recognized by cells of the adaptive immune system. They are one of the bridges between the innate and the adaptive immune systems.

Two key members of the adaptive immune system are types of white blood cells called lymphocytes. The first is the B cell, which is born in the bone marrow and travels while still immature to various lymphoid tissues such as the spleen, the lymph nodes, and the immunological tissues of the gut wall. As it matures, it will become capable of producing receptor molecules on its surface in response to the detected presence of antigens on invading microorganisms. These receptors are immunoglobulin molecules with a tip that is hypervariable. The genes responsible for this hypervariable region can be rapidly mutated to produce an almost infinite number of combinations, so an exact fit can be made to bind to any specific antigen protein on a pathogen. Thus, naive B cells can rapidly produce the correct lock for any antigen key, bind to it, and neutralize it. At this point the B cell either transmogrifies into a plasma cell and becomes a factory for that particular antibody, churning out millions of copies that free-float in the blood ready to bind to more antigen, or it becomes a memory B cell, which can survive for long periods in the body, remembering the antigen that activated it, and ready to immediately go into action whenever that antigen is sensed again.

The second key member of the adaptive immune system is the T cell or T lymphocyte. Its precursor travels from the bone marrow to the thymus gland (hence the T) to complete several further stages of maturation. One group of T cells, the effector T cells, present receptors on their cell membranes that are also hypervariable and can be manufactured to recognize any antigen complex detected on the coat of invading viruses and bacteria. They do not release antibodies into the blood but directly attack the invaders and destroy them. They can rapidly expand their population number to destroy all invaders bearing the same antigen signature, and a number of them linger on in the blood and lymph for as long as twenty years, a semipermanent memory. It is this adaptive immune system memory that gives us natural immunity in which subsequent infections with the same pathogen become milder because the memory B cells and T cells are like a gun already cocked. It is also the basis for vaccination, which introduces dead, inactivated, or attenuated pathogens, or the antigens stripped from the outer coat of one specific pathogenic organism, to establish permanent clones of memory lymphocytes, which can leap into action should the real pathogen arrive on the scene. So, adaptive immunity recognizes specific antigens, produces specific receptors and antibodies to fight them, remembers any specific pathogen, and rapidly deploys should it reappear. Both the B cells and the T cells go through an instruction process either in the bone marrow or thymus that weeds out those cells that react too strongly to "self" antigens, the protein markers for the host body's own cells. Thus, in the main, they are "taught" to tell the difference between "self" and "non-self" and act accordingly.

Crucial to an understanding of allergy and autoimmunity are two other types of T lymphocytes. First are the T helper cells (Th cells, for short).

They are so called because they assist or help other white blood cells to mount a challenge to pathogens. Once activated by antigens "presented" by other cells, they produce cytokine-signaling molecules that can either damp down or inflame the immune response. The three types of Th cells we are concerned with are the Th1 cells, which are implicated in autoimmune diseases; the Th2 cells, which are implicated in the response to intestinal worms and common allergens (and are therefore implicated in allergy and all the atopic illnesses); and the Th17 cells, which are very potent defenders against a range of microbial invaders and whose cytokine products can be highly inflammatory—as such they are also frequently associated with a range of autoimmune diseases. Finally, we have the regulatory T cells (Tregs). Their job is to prevent the Th1 or Th2 cells, and other lethal T cells, from getting out of control by suppressing or regulating their activity. This ensures that the inflamed, cytotoxic immune response can be reined in once any particular pathogen has been dealt with.

In the early days of the hygiene hypothesis, it was assumed that Th1 cells and Th2 cells were antagonistic, operating rather like a seesaw such that when Th1 cell populations were high, they inhibited the production of Th2 cells, thus preventing allergies. Conversely, it was thought that if the seesaw tipped the other way, so that Th2 populations were high, Th1 populations would be depressed and so autoimmunity would be prevented. However, it soon became obvious that in some autoimmune conditions, patients were also atopic (like the Johnson family) and that allergies and autoimmune disorders were simultaneously rising in all Western countries. This is what has led to a reinterpretation of immune system dynamics, and it is now believed that it is the regulatory T cells that operate as a master switch to turn down all effector T cells—both Th1 and Th2—in the immune system. "Old friends" organisms prevent our immune systems from treating them with suspicion and attacking them because they stimulate the production of Treg cells, reducing the attack squadrons of effector T cells, and thus bringing about a state of immune tolerance. For instance, June Round and Sarkis Mazmanian have investigated the important probiotic bacterium *Bacteroides fragilis*, which is prominent in the gut of most mammals. They have shown that *B. fragilis* produces a specific symbiotic molecule called polysaccharide A (PSA), which signals directly onto one of the main receptors of regulatory T cells. If you remove PSA from the *B. fragilis*, it is immediately unmasked and Th17 cells move in and prevent it from colonizing the gut.

There is a general principle at work here. A wide range of microbes and fungi needed to be tolerated by human immune systems because they were ubiquitous, over millions of years, in food and water, and consequently infected us. Similarly, helminthic parasites needed to be tolerated because, although not always harmless, once they became established in us, they were difficult to dislodge and an immunological attack on them would cause disproportionate collateral damage. For instance, sustained attempts by the immune system to destroy the larvae of the nematode parasite *Brugia malayi* cause lymphatic blockage and elephantiasis. Over millennia, a state of interdependence seems to have evolved. These commensal organisms, sharing the dining table of our guts, needed to regulate our immune systems so that they could live inside us without being attacked, and immune regulation became a necessity for us to prevent our immune systems from overreacting to these long-term residents in such a way that we self-harm. But this has meant that, in a sense, we have handed the control of our immune systems over to the microbiota inside us. The downside is that while immune regulation works perfectly well when we have a rich assortment of "friendly" bacteria, fungi, and worms in our guts, things start unraveling very fast when they are no longer there. Our powerful immune systems, evolved to work in the presence of relatively harmless endoparasites, lose their brakes and won't switch off, resulting in the chronic inflammation that leads to the modern plagues of allergic and autoimmune illness.

Matteo Fumagalli, now at the University of California, Berkeley, hypothesized that parasitic worms would have exerted very significant selection pressure on humans throughout their history. Even today, he says, over 2 billion humans are infected with parasitic worms that cause widespread childhood morbidity. They stunt growth, leave individuals open to other infections, and cause preterm births, low birth weights, and maternal mortality. It was ever thus. Fumagalli argues that the selection pressure exerted by worms would have been much greater than that exerted by bacteria and viruses or climate, and that it should be possible to see it reflected in our genomes, particularly among genes that are implicated in immune responses. Using data from the Human Genome Diversity Project, he sampled 950 individuals from all over the world and correlated the occurrence of gene mutations with species diversity of helminths in the area of the world from which each individual came. Nearly one-third had at least one gene mutation significantly associated with helminth diversity, and in all they counted over eight hundred gene mutations. Many of these genes were involved in regulatory T cell function and in the activation of the macrophages of the innate immune system. Others were involved in the cytokines produced by the Th2 cells that are mobilized against helminth infection.

Importantly, the gene variants they found give us a huge clue into the nature of the love-hate relationship between humans and helminths. While many of the genes that had evolved under pressure from helminths were associated with aggressive, pro-inflammatory responses geared to fighting helminth infections, other genes worked in the opposite direction by stimulating immune tolerance through regulatory T cells, anti-inflammatory cytokines, and other molecules that inhibit immune responses. Helminths are the masters of immune regulation. This is why they can persist in any host for a long time and persist as major human infections over millennia. Graham Rook sees the relationship between them and us as a game of chess, or as a dynamic tension between parasite and host immunity. He suggests that where helminth load was particularly high, there would have been selection for more pro-inflammatory gene variants either to counter the powerful immunoregulatory action of the helminths or to keep the immune system effective against other viral and bacterial insults in the face of helminthic regulation. This dynamic equilibrium is shattered as soon as the helminths are eradicated, leading to the inflammatory overshoot we see today in allergic and autoimmune disease.

Jim Turk is a biological security officer on the University of Wisconsin campus. It's his job to make sure all the laboratories handling pathogenic or recombinant organisms are safe and contained. Being in good shape had always been important to him—he used to run marathons. Then, in spring 2005, his speech suddenly became slurred. His wife feared he had had a stroke and made him go to the doctor, who ran some tests but could find nothing wrong. "You're overdoing it," the doctor said. "You're working full-time, you're going to university all the time, you've got a young family. You're tired and stressed." Jim noticed occasional balance problems and some numbness in his lower legs, but, because of his doctor's assurances, he brushed them aside. Then, in February 2008, he began serious indoor training for another marathon. After three or four minutes of hard running in the hot arena, he remembers, "I would lose control. I would be grabbing at railings. I'd be stumbling around. Of course being kind of dumb or in denial, I decided I was going to keep going back. 'I must be in worse shape than I thought,' I told myself. And so I went back the next day and the next day, and it was always the same thing until one day I fell flat on my face."

Still in denial, he busily set about coaching his son's baseball team only to notice that he was moving his legs wider and wider apart just to balance himself upright on the baseline. Bending down sent his head spinning. Back to the doctor he went. High blood pressure was quickly ruled out—he had the heart of a teenager—and then a neurology referral and an MRI scan made everything horribly clear. "My brain was peppered with these plaques. I had a lot of them—twenty or so." Another MRI scan showed that the plaques extended right down into his spine, and he was diagnosed with relapsing-remitting (early stage) multiple sclerosis in August 2008.

Jim had joined over 2 million people worldwide who have MS. They mostly live in Westernized countries and tend to be young—mean age of onset is twenty-five. Although over fifty genes, lack of sunlight, vitamin D, viruses, smoking, and affluence have all been implicated, the strong inverse correlation between helminth infections and MS is striking, as is the fact that T cells are often dysregulated in people with MS. Each MS plaque is a small area of inflammation and tissue damage where the myelin sheath that coats and insulates the nerve fibers is destroyed. In relapsing-remitting MS, there are typically five to ten new plaques formed every year, though only one in ten, on average, will hit a critical area of the central nervous system like the optic nerve, cerebellum, or tracts of sensory nerve fibers.

A few nights after his diagnosis, Jim and his wife were watching a weekly feature on multiple sclerosis on their local television channel: "Dr. Fleming, from the University of Wisconsin, was on and he was talking about his new HINT (helminth-induced immunomodulatory therapy) trial using whipworm eggs. I already knew a little bit about the hygiene hypothesis, so I was immediately intrigued but, having a science background, realized it might be tough to get people to line up to swallow whipworm eggs just because of the way it sounds. I knew you wouldn't be able to see anything—the eggs are microscopic. But if you tell people they're drinking worm eggs . . ." Jim contacted Fleming's group the next day and became the first volunteer to be enrolled in the study. He took the worm eggs for three months and then had a sixty-day washout period.

John Fleming's interest in the potential of *Trichuris suis* had been spiked by the success of Joel Weinstock and his colleagues with inflammatory bowel diseases, where, as we discovered earlier, some 79 percent of patients who received the whipworm got better in a stage 1 trial. Although the pig whipworm eggs do not go through their full life-cycle in humans (they may not even hatch) and they last only a few weeks at most before being eliminated in feces, successive treatments seem to give the whipworm enough transient presence in the gut to modulate the immune system. Weinstock believes they may do this directly, by secreting immune-regulating molecules, or indirectly, by affecting the composition of the gut microbiota, which then does the job. Fleming says that the evidence for a failure in immune regulation being at the heart of MS is strong. Such regulation, he explains, is achieved by interaction between regulatory T cells, dendritic cells, regulatory B cells, cytokines, and other effector cell types. All can be disordered in MS sufferers, and over twenty mechanisms by which helminths influence immune regulation have so far been discovered.

Fleming had also taken notice of Jorge Correale's work in Argentina that reported much reduced severity in symptoms of MS in those patients with an intestinal worm infection. Correale had also shown that the worm effect was very specific to MS because the regulatory T cells produced in worm-infested patients were specific to myelin peptide, meaning they were protective of the myelin sheathing around nerves. Fleming obtained permission for the small initial phase 1 trial that involved only five newly diagnosed patients who had yet to receive any form of treatment for MS. Although the trial was deliberately short-lived and testing mainly for safety rather than efficacy, Fleming noticed a substantial reduction in new brain lesions in all five volunteers while they were taking the whipworm eggs, whereas the lesions increased in number once treatment had stopped.

Jim's MS is now being treated with conventional medication, and he tries to keep his illness at bay through a combination of careful diet and exercise. "In fact, if I didn't have MS, I'd probably be in the best shape of my life!" He has learned to contain his symptoms, especially his speech impediment, so well that most of his colleagues are totally unaware of his condition. But playing sports with the kids is a thing of the past. "As my kids grow up—they're both boys, thirteen and nine—I'd like to be able to play football in the backyard with them, shoot baskets, or go for long bike rides. I can still throw the ball a little—I just can't run with it. And I'd love to be able to go out for runs. I haven't been able to do that in six years. That's what I wish I could do the most."

Our resident gut microbiota—the mass of over two thousand bacterial species identified as frequent, long-term inhabitants inside us—is extremely complex. Our relationship with them is so close and intertwined that many of the metabolic signatures that can be identified in human blood, sweat, and urine actually come from our commensal bacteria, not us. When we look at any individual's response to a particular drug treatment, it may be that what we are looking at is the reaction of our gut microbe colonies, not our own bodies. Only vertebrates have such rich, enduring microbial colonies. All invertebrates have tiny gut colonies, sometimes only several species, and these are often transient. There is another interesting difference between invertebrates and vertebrates. Invertebrates do not have an adaptive immune system; they only have the more primitive innate immune system.

That observation has prompted Margaret McFall-Ngai, a medical microbiologist from the University of Wisconsin, to stand modern immunology on its head. She argues that the adaptive immune system evolved not only to protect us against environmental pathogens, but also to police the resident microbial community inside us. The pioneering microbiological research of the nineteenth century, for which we have to thank Koch and Pasteur, was done in the context of human disease. It was set on its course only to view microbes as invaders of the human body, capable of causing infection. Pathogenic bacteria and viruses can mutate far faster than we can, and rapidly change the antigenic markers on their outside coat by which our immune system recognizes them as the enemy. We needed, states conventional immunology, an adaptive immune system with long-term memory and the ability to generate an almost infinite variety of matching antibodies to counter them.

While it is true that we could not effectively counter pathogenic infections without an adaptive immune system, it is also true, says McFall-Ngai, that the complexity of the pathogenic environment is dwarfed by the complexity of the microbial communities within us. To put this in perspective, it has been estimated that only twenty-five infectious diseases account for the vast majority of human death and disability, and ten of these could only have arisen after urbanization some six thousand years ago because they depend on human-to-human transmission. They could not have survived when humans existed in small, widely dispersed, roaming communities before the advent of agriculture and animal husbandry. Our gut microbiota can consist of thousands of species, and, although they may be friendly most of the time, bacteria are a perfidious bunch; they are opportunistic and can easily mutate to turn from being benign to being pathogenic when it suits them, for instance, when damage to the gut wall makes it leaky and they can no longer be safely corralled in the gut lumen.

So, not only has our microbiota been around far longer than most pathogens, but it is far richer in number and species diversity than the pathogenic environment. Without it, the "gut police" of our adaptive immune systems simply would not develop into a highly plastic system capable of discriminating between friendly microbes and the pathogens lurking among them, or detecting friendly bacteria who have turned rogue. According to Sarkis Mazmanian, our microbiota presents a challenge to the adaptive immune system because it contains an enormous foreign antigenic burden, which must be either ignored or tolerated to maintain health. In turn, it is in our microbiota's self-interest to maintain the health of its host. It is humbling to remind ourselves that we are simply an attractive niche or home for these microbes—one that they have fashioned to suit themselves. We, and our microbiota, have coevolved to work together to repel pathogens, because it is in our mutual interest. For example, it has recently been shown that mice in the throes of a systemic bacterial infection switch on production of a sugar that specifically favors the population growth of friendly bacteria in their guts, which rally round to help repel ensuing infections.

Gérard Eberl, of the Institut Pasteur, believes that in this superorganism of human and microbes, the immune system is never at rest; it is like a spring. The more the microbes colonize the niches within us or behave like pathogens, the stronger they pull the spring of immunity, he says, and that stronger spring of immunity pushes the microbes back. The immune system is therefore always under tension, the tension required to maintain homeostasis—the status quo. For instance, if we take away our gut microbiota with antibiotics, we can become susceptible to infection by *Enterococcus*. This is because it is the friendly bacteria that make our gut wall produce the toxic antibacterial peptides that normally fight this pathogen. Too weak an immune system, explains Eberl, leaves the superorganism vulnerable to old friends going bad and turning into opportunistic pathogens; too strong an immune system destabilizes our microbiota, and we progress into autoimmune disease.

McFall-Ngai says it is high time that we started viewing the microbes inside us as a whole organ, similar to but much more complex than the heart, liver, or kidney. In its complexity, it more closely resembles the brain. The brain is composed of over 80 billion neurons, massively interconnected; the microbiota is composed of over 80 trillion organisms massively communicating with one another via signaling molecules. Both have a memory, both can learn from experience, and both can anticipate future uncertainties. The gut has been called the "second brain" and has its own dedicated nervous system embedded throughout the gut wall. It is becoming increasingly clear that our gut microbes can communicate directly with our brains, and they are implicated in brain development, brain chemistry, behavior, and mental illness. They produce hundreds of neurochemicals, including most of the body's supply of serotonin, and there is two-way communication such that the mix of bacterial species in the gut can influence the brain and vice versa.

Much of the research that demonstrates this uses mice as the experimental model. For instance, Premysl Bercik has compared two strains of mice—one timid, one bold and adventurous. A group of animals from each strain were cultivated germ-free. The germ-free timid mice then had their guts inoculated with the gut contents of bold mice that had been reared normally. Germ-free mice from the bold strain were conversely inoculated with the gut contents of normally reared timid mice. Their behavior immediately switched. Timid mice became bold and bold mice, timid. John Bienenstock gave a timid strain of mice a broth heavily laced with a popular probiotic bacterium, *Lactobacillus rhamnosus*. After twenty-eight days, they were more willing to enter a maze than controls and less likely to give up and float on a forced swim test. The activity of stress hormones in the brain was reduced. Similarly, rearing germ-free mice in a confined area stresses them and raises the activity of the hypothalamus-pituitary-adrenal axis. This pushes up levels of corticosterone and adrenocorticotrophin, two stress hormones. But all this activity subsided after the mice were inoculated with another common probiotic, *Bifidobacterium infantis*. The reverse also happens. Michael Bailey has found that macaque infants from mothers who were stressed with loud noises during pregnancy had lower counts of friendly bacteria like *Lactobacillus* and *Bifidobacterium* in their guts, while another researcher found fewer lactobacilli in the stools of college students during exam week.

How can the bugs in our guts "talk" to the brain and vice versa? What is the link? In a very recent paper, Emeran Mayer and Kirsten Tillisch report on their fMRI brain-imaging study of the effects of probiotic bacteria on mood and brain activity in a group of healthy, normal female human volunteers. One group of women was untreated, while the other group was given fermented probiotic drinking yogurt twice a day for four weeks. They were imaged before and after treatment: while they were undertaking a task where they had to look at a range of emotional faces and while simply resting. The researchers identified a pathway into the brain involving a tract of nerve fibers in the brain stem called the nucleus tractus solitarius. This receives inputs from the vagus nerve that innervates the gut. From this brainstem nucleus, circuits were activated that ramified into higher brain centers, including the amygdala (the fear or emotional center of the brain), the insula, and the anterior cingulate cortex, all regions that are involved in processing emotional information. In those volunteers who had taken the yogurt, there was reduced activity in these circuits, suggesting decreased arousal or anxiety associated with the tasks. These women had calmer emotional reactions. Although the results should be interpreted with extreme caution, a reasonable working hypothesis is that the probiotic bacteria in the gut were able to signal to the brain using the vagus nerve — literally communicating gut feelings.

A recent paper by Joe Alcock, Carlo Maley, and Athena Aktipis has pulled together many sources of evidence suggesting that the bugs in our guts can influence what we eat, producing cravings for foods that will give them a competitive edge in the large intestine and inducing profound states of unease or dissatisfaction in us until we are sensually rewarded by particular foodstuffs, like chocolate, that will equally reward the bacteria with the food they hanker after. Our gut bacteria are using the vagus nerve to manipulate our behavior. This raises the intriguing possibility that by changing the species composition of our microbiota, we might be able to change eating habits and perhaps even ward off obesity.

What happens when the brain malfunctions? Inflammation and gut pathology are both involved in autism, for instance. Children with autism frequently have evidence of inflammation in their brains, and there is growing evidence that this inflammatory state is transmitted from their mothers during pregnancy. Alan Brown, professor of clinical psychiatry at Columbia University, has studied data from the Finnish Cohort Study, where 1.6 million blood samples were collected from over 800,000 women during their pregnancy. They matched the levels of C-reactive protein (CRP; a blood indicator of inflammation) with the risk of autism in their children. The risk was increased by 43 percent among mothers in the top twentieth percentile of CRP levels, and by 80 percent among women in the top tenth percentile. Since CRP is also elevated in the immune response to infections, this suggests that infections during pregnancy in some mothers, as well as inflammation caused by autoimmunity, can communicate an inflammatory state to their fetuses through the placenta. There is further evidence from Danish population-wide studies that autism risk is raised by 350 percent in babies born to mothers with celiac disease, and 80 percent in mothers who suffer from rheumatoid arthritis. Eric Hollander, the autism expert who treats Lawrence Johnson, suggests that something as simple as influenza in the mother-to-be can feed-forward an inflammatory state from mother to fetus via pro-inflammatory cytokines. These are similar responses to those triggered in fetuses when their mothers have lupus, an inflammatory autoimmune disorder that causes fever, swollen joints, and skin rashes.

Autistic children may also inherit, in the genetic sense, a hyperactive immune system that leaves them vulnerable to developing autoimmune disease. About 70 percent of children with autistic spectrum disorder have severe bowel irritation. They can suffer from diarrhea and painful abdominal distention, and this is related to their irritability, aggression, and self-harm. Endoscopic examination often reveals an inflammatory pathology very similar to Crohn's disease and ulcerative colitis. Stephen Walker, from the Wake Forest Institute, compared patterns of gene expression in biopsy material collected from the guts of autistic children with irritated bowels and adults with IBD. While there were significant differences in the pattern of genes affected, there was also significant overlap in a number of genes that were either turned up or tuned down in both conditions. This suggests autistic children with irritated guts, and non-autistic adults with bowel disease, both suffer from autoimmunity. Mouse models of autism show that helper T cells are made permanently hyperresponsive by maternal infection during pregnancy, and regulatory T cells are reduced in number.

What about depression? It cannot be said too clearly that depression is not an inflammatory disorder per se. Many patients presenting with depression do not have strong evidence of inflammation, and many individuals have high levels of inflammation markers in their blood without becoming depressed. Having said that, there is clearly a major subgroup of individuals who are more prone to reacting to a background inflamed state by becoming depressed. There is a fascinating example of the interplay between inflammation and depression in the effect of interferon-alpha (IFN-α)—which is a potent inflammatory cytokine—when it is used as a therapy for hepatitis C or cancer. At high doses, fully 50 percent of patients will develop major depression within three months of commencing therapy. Downstream, IFN-α induces a cascade of other inflammatory cytokines, like interleukin-6 and tumor necrosis factor-alpha (TNF-α), which also correlate with depression. But is the converse true? If you remove the inflammatory cytokines, does the depression lift? One study examined patients who were being treated for Crohn's disease with a monoclonal antibody called infliximab. A proportion of them were also clinically depressed. The infliximab removed depressive symptoms but only in those individuals who had high levels of C-reactive protein in the bloodstream, indicating a state of high inflammation. This was because infliximab is a potent TNF-α agonist, so it is likely that while curing the Crohn's disease, it was also neutralizing the cytokine that stimulated the inflammation and therefore alleviated the depression.

In susceptible individuals, depression is only one of a number of disorders that seem to be caused by chronic low-grade inflammation. These include cardiovascular disease, stroke, diabetes, cancer, and dementia. Moderate increases in levels of chronic inflammation are enough to predict the future development of all these modern disease states in presently asymptomatic individuals. In the Whitehall study of UK civil servants, circulating levels of C-reactive protein and interleukin-6 were inversely correlated with employment grade, implying that the lower the pecking order, the higher the background inflammation became. The psychologist Andrew Steptoe used this gradient to successfully predict occurrence of depression twelve years down the line. Further studies show that depressed individuals with histories of early life trauma or neglect release more interleukin-6 in response to stress tests. It could be that it is not simply the stress of modern Westernized lifestyles and workplaces that causes inflammation and thence depression, but a relative lack of immune regulation in Westernized societies that allows inflammatory cytokines to run amok in reaction to them. In which case, how, specifically, could the "old friends" hypothesis come into play in this scenario?

Tom McDade, of Northwestern University, has been comparing populations in developing countries with US populations in an attempt to disentangle infection, inflammation, stress, depression, and morbidity. He notes that levels of C-reactive protein are transiently high in a tribe of Amazonian Ecuadorian Indians and correspond to frequent bouts of infection. But as soon as the infection subsides, so do the CRP levels. Their profile is a series of peaks and troughs, whereas in the United States, levels of CRP are stable and high even in the absence of high rates of infectious disease. This chronic, persistent inflammation indicates poor immunoregulation. McDade looked at a rural population in Cebu, in the Philippines. He measured the levels of microbial diversity in and around each village house by examining animal feces and measured the frequency of childhood diarrhea and the numbers of births during the dry season when infectious loads were highest. All these factors predicted low CRP levels in adulthood and reflected early high microbial exposure.

McDade then looked at the effects on children of separation. All children, as you would expect, were distressed at losing their mothers, but that stress did not raise their CRP levels as long as they came from a home typified by a high level of microbial diversity. It did not even rise if they subjectively felt psychologically disturbed by the pain of separation. So, in rural Philippines, thanks to adequate childhood exposure to "old friends" microorganisms, temporary depression, social stress, and unhappiness never led to damaging inflammation. McDade also found that concentrations of the pro-inflammatory cytokine interleukin-6 were generally low in Philippine populations, while levels of the anti-inflammatory cytokine interleukin-10 were exceptionally high. Obese women in the United States generally have high levels of IL-6, but in the Philippines, women with similar waistlines don't. When they looked at men with high skin-fold thickness in the Philippines, it was not associated with high CRP levels, whereas this is exactly what you expect to find in the United States. There was, he concluded, protection against inflammation on all fronts in which the early microbial environment of upbringing seemed key.

We may be entering an era where microbiology and immunology, specifically the "old friends" hypothesis, begin to make a real impact on public health policy. For instance, Martin Blaser is deeply worried about the overuse of antibiotics. We are all aware of the increasing dangers of multiple antibiotic resistance because it is giving rise to a race of dangerous "super-bugs" that are becoming almost impossible to treat. But regular broad-spectrum antibiotic treatment is also killing off the friendly and useful commensal bacteria inside us with disastrous results. By the time they reach the age of eighteen, Blaser points out, American children, on average, will have received between ten and twenty courses of antibiotics, killing off friend and foe alike. In some cases, he says, our microbiota never recovers and we are fueling the dramatic increases we see in type 1 diabetes, obesity, inflammatory bowel disease, allergies, and asthma. Occurrence of IBD, for instance, rises with the number of antibiotic courses taken. Worse still is the industrial-scale administration of antibiotics to farm animals purely to assist them to put on weight. Antibiotics are routinely given to nearly one-half of all women in pregnancy in the United States, and since babies acquire their gut microbes from their mothers, each generation could be beginning life with a smaller endowment of friendly microbes than the last, and so on *ad calamitas*.

A scary scenario of what that calamity might be has recently arrived from Sven Pettersson, of the Karolinska Institutet in Stockholm. It is known, explains Pettersson, that there is an intestinal barrier to prevent the trillions of bugs in our guts from escaping into the body. That barrier is actually created and maintained by the friendly bacteria inside us. In experiments on mice, he has discovered that the gut microbiota exert a similar control on the impermeable blood-brain barrier that protects the brain from insult by a huge variety of molecules and microorganisms. Baby mice born from germ-free mothers had leakier blood-brain barriers that persisted throughout life. Although this research has yet to be transferred to humans, the implications are extremely worrying. If a depleted gut microbiota in the mother can lead to a defective blood-brain barrier in the baby, then the proper development of the brain, and later protection of the brain, might become heavily compromised. This may cause us to become much more wary of routine antibiotic treatment of pregnant mothers and Cesarean section for the delivery of their babies, because we already know that both interventions deplete the gut microbiota that babies inherit from their mothers.

A massive amount of research now amply demonstrates the degree to which a benign microbiota inside us can protect our health. But one researcher believes that the beneficial effect of microbes can extend to both the urban or rural environment around us. Ilkka Hanski, from the University of Helsinki, thinks that microbiology—specifically the "old friends" hypothesis—should be taken into account in town planning, particularly in considering the importance of green spaces. He has recently reported a significant relationship between skin allergies, vegetation, and land use in a heterogeneous group of 118 Finnish teenagers chosen to represent a range of living environments from towns to villages to farms. He took skin swabs to look at the diversity of skin bacteria, a skin allergy test to check levels of atopy, and measured land use and plant cover in the immediate vicinity of their houses and up to three kilometers away. He found a strong relationship between atopy and a group of bacteria called gammaproteobacteria, which were significantly less diverse in atopic individuals. He then went on to measure levels of the anti-inflammatory cytokine interleukin-10 in the blood and found that one gammaproteobacterium, *Acinetobacter*, was strongly linked to high levels of IL-10 in the blood of healthy individuals, but not in allergic individuals.

These "protective" bacteria are commonly found on plants and pollen, as well as in soil, which is why Hanski found such a strong association between diversity of skin bacteria, lack of atopy, and the richness of the surrounding vegetation, particularly in the less common species of flowering plants. The teenagers would have picked up the bacteria through contact with soil and vegetation, or through pollen or wind transfer. More and more of us worldwide are moving to cities where open green spaces may be few in number or completely absent. If we are dependent upon certain bacteria to encourage high levels of anti-inflammatory cytokines in our blood, thereby inducing immune tolerance, and if those bacteria, in turn, are dependent upon the richness of vegetation, then the importance of green spaces goes far beyond a feel-good factor and is at the root of allergic conditions and public health in general, where, as Hanski notes, they may have profound consequences.

In a similar vein, we have seen how Mikael Knip's Karelian study has identified the microbiota to be an important factor in type 1 diabetes. Russian Karelian children not only had a more diverse microbiota but higher levels of regulatory T cells in their blood. His continuing research hopes to more specifically identify the species of microbe most important for protection against autoimmune and allergic disease so that he can design a medical intervention in the near future for children at risk. Other allied metabolic disorders steeply on the rise, like type 2 diabetes and the obesity pandemic, have also been shown to relate to immune dysregulation and loss of microbial diversity in the gut.

There is a huge and growing literature on the effects of negative life events and loneliness or social isolation on our future health. Although many of these long-term epidemiological surveys show correlations between inflammation, stress, social isolation, and socioeconomic status, only one or two to date have looked at gut microbial diversity. Graham Rook, however, would bet the farm that, had they done so, they would have found that those individuals who showed poor resistance to the vicissitudes of Western life and had chronic high levels of inflammation, and mental or physical morbidity, would typically show reduced diversity in their gut microbiota and consequently compromised immunoregulation. The "old friends" hypothesis, thinks Rook, could be the "missing variable" in all these public health studies.

A few examples will suffice to see how wide-ranging these implications could be because we are clearly looking at a cradle-to-grave phenomenon. Per Gustafsson, for instance, has examined the way that isolation and subjective feelings of unpopularity at school adversely affect health several decades down the line, correlating with psychiatric problems, cardiovascular problems, and diabetes. Gregory Miller and Steve Cole have gone further and specifically linked childhood stress and adversity to chronic low-grade inflammation. Using data from a large group of Vancouver adolescents, they show that depression and inflammation (as measured by high circulating levels of C-reactive protein and interleukin-6) co-occur, but only in those individuals who had suffered childhood adversity. CRP levels linger on after the depressive episode has ended, which may make such children vulnerable in the long term to persistent mood disorders, cardiovascular disease, diabetes, and autoimmunity, they say.

Bruce McEwan introduces the idea of "allostatic load"—the cumulative wear and tear on the organism as it tries to adapt to life's demands. They show how the stresses associated with gradients of socioeconomic status show up as chronic inflammation that upsets neuroendocrine function and can lead to heart problems, osteoporosis, metabolic disorders like diabetes, and cognitive decline. W. Thomas Boyce and Kathleen Ziol-Guest are even more explicit about the relationship between childhood poverty and adult disease. Although factors like diet and nutrition are also important, the chronic inflammatory processes set running by childhood exposure to negative life events are an additional common pathway to several chronic morbidities, they explain. Chronic inflammation is controlled by events in the brain, interplay between the hypothalamus, pituitary gland, and adrenal gland, and the cellular immune system where T lymphocytes are encouraged to differentiate into Th1 and Th2 cells, which can result in widespread tissue damage if inflammatory conditions persist. Low childhood socioeconomic status is associated with higher blood levels of C-reactive protein, cytokine IL-6, and the other pro-inflammatory cytokine, TNF-α, placing such children at greater risk of developing inflammatory diseases such as atherosclerosis, autoimmune disorders, and cancer. This is precisely where Tom McDade's research enters the picture by measuring these early environmental stressors, specifically the loss of the mother, in Filipino children and showing that the crucial protective factor against early insults that might normally lead to later metabolic and mental illness was microbial exposure in infancy.

We shouldn't really need the "old friends" hypothesis to tell us that we ought to provide better quality care for our aged population in residential homes. But a recent survey by Marcus Claesson and his colleagues at the University of Cork graphically shows how toxic the environment of care homes can be to its residents. Claesson identified 178 elderly people, with a mean age of 78, in southern Ireland and divided them into those who were still living in the community, those who were in a hospital for short-term rehabilitation, and those in long-term residential care. He took fecal microbial samples, dietary information, and measures of their immune status. He found that the gut microbiota in those individuals in residential care was much less diverse than in those living in a community setting, and this correlated with high and persistent inflammation and increased frailty. As we get old, our teeth don't work so well, we produce less saliva, our digestion suffers, and we get more constipated. All this damages our gut microbes. The combination of a poor, bland diet (and probably social isolation) in nursing homes simply makes the situation much worse, taking its toll on the microbiota, driving chronic inflammation, accelerating aging, and deteriorating health. Given a rapidly expanding aged population in Western countries, says Claesson, dietary interventions to prevent this accelerated morbidity and premature death should become a priority.

Across the world, researchers are trying to turn many of the microorganisms identified by the "old friends" hypothesis into mainstream medicine. You can tell when a new applied science is in its infancy because it is often populated with self-experimenting pioneers, people who try techniques out on themselves before subjecting patients to them, or people who venture into unproven therapies out of desperation. People like David Pritchard, from the University of Nottingham, who, while doing field research in Indonesia, became intrigued by the observation that individuals infected with hookworm seemed protected against allergic diseases. He later deliberately infected himself with hookworm larvae, via a scratch in the skin, in order to prove to his satisfaction that the negative effects of hookworm infection were tolerable compared with the hoped-for benefits in immunoregulation. He is now connected with a major medical trial at Nottingham that aims to see if hookworms can moderate disease progression in multiple sclerosis.

In 2004 a young man, who will remain anonymous, traveled to Thailand to deliberately infect himself with human whipworm eggs, procured from the feces of an infected girl, to see if they could cure an ulcerative colitis that was so resistant to cyclosporine treatment that he was in immediate danger of having his entire colon removed and replaced with a colostomy bag. He had previously approached Joel Weinstock for treatment with *Trichuris suis*, but Weinstock had to refuse him on ethical grounds. Within three months of self-treatment, bowels that had once produced over a dozen bloody movements a day were back to normal. It appeared that the worms were inducing high amounts of interleukin-22, which is important for healing the gut mucosa. He is now one of a number of volunteers helping P'ng Loke, at New York University, to investigate the effects of helminths on inflammatory bowel disease.

The road from bright biological insights to tried-and-tested pharmaceuticals is often a rocky one. In order for any new drug to reach the market, it has to be rigorously tested in large, randomized, double-blind trials that are capable of accounting for any placebo effect and dissociating it from any true efficacy of the drug being tested. By this measure, the recent slew of medical trials for pig whipworm eggs has so far proved disappointing. A large trial of whipworm eggs among sufferers of Crohn's disease was recently halted because it could find no efficacy, and John Fleming's phase II trial of whipworm eggs for multiple sclerosis has also failed. The problem may be because the whipworm species chosen for both these trials is not well suited to human beings. *Trichuris suis* is a parasite of pigs—not humans. It may not even multiply in the human gut, and so the "infection" is flushed out in the feces within a couple of weeks. This is why Stewart Johnson's son has to continually take eggs to stand a chance of the treatment working. Because humans are not the natural home of *Trichuris suis*, the eggs may not be regulating human immune systems as effectively as would a human-specific helminth species.

We may, says Graham Rook, the pioneer of the "old friends" hypothesis, be using the wrong worm on the wrong people. The reason why Jorge Correale finds that tapeworm infection causes remission in symptoms among his multiple sclerosis sufferers in Argentina may be because, due to the endemic nature of tapeworm in parts of South America, they had asymptomatic tapeworm infections during early childhood when their immune systems were maturing. Their immune systems will have become calibrated by this previous early insult thanks to developmental or epigenetic effects by which the helminths regulate key immune system genes. This previous exposure would not be shared by Crohn's disease and multiple sclerosis sufferers in North America.

Meanwhile, Eric Hollander *has* achieved modest success with a small trial of *Trichuris suis* in adult patients with autism. There were reductions in scores on several psychological tests that measure autistic symptoms—reductions that hovered just below statistical significance. They were less likely to have temper tantrums or "act out," says Hollander, and were less compulsive and more tolerant of change. Hollander is now conducting a larger trial of *Trichuris* with children and younger adults with autism.

Whatever the final results from Hollander, because of the acknowledged link between autism, poorly regulated immune systems, and gastrointestinal discomfort, Rosa Krajmalnik-Brown, at Arizona State University, examined the microbiota of autistic children and found them to be generally less diverse and lacking several important "friendly" bacterial species compared with the gut microbiota of normal children. She is currently running a trial with autistic children where she introduces fecal transplants taken from normal individuals to repopulate their guts with a diverse and protective microbiota. Fecal transplants could be a more effective way of rapidly repopulating the gut than orally taken probiotics, which may have a limited effect because when you take a few pills of probiotics or drink a glass of fermented probiotic yogurt, you are trying to improve the gut microbiota by introducing a few hundred million microorganisms by mouth into a lower gut population of scores of trillion. Also, common probiotic products abound with lactobacilli and bifidobacteria that, while vital in establishing infant immune systems, are only minority players in adult guts. We need new probiotics.

All these "old friends" researchers look forward to the near future when their efforts may lead to a "new pharma." Helminth researchers realize that we cannot go on forever feeding whipworm eggs to people or introducing hookworm larvae under the skin. Sarkis Mazmanian has shown the way through his work with the bacterium *Bacteroides fragilis*. He found that he could dispense with the whole intact bacterium because the polysaccharide A molecules extracted from them were capable of regulating the immune system. In a similar fashion, Joel Weinstock continues to research the mechanisms that helminths use to evade our immune systems. He hopes, in the future, to be able to isolate the molecules they use to do this and thus lay the foundation of a new evolution-inspired drug therapy for this range of illnesses.

This approach would greatly please Stewart Johnson, whose heroic, and ultimately successful, research saved his son Lawrence from the twilight world of a lifetime in institutionalized care. As Stewart says: "I did not see this coming. I did not expect this to work. I was just satisfying the scientist in me. To try and never stop. Until I die, I'm going to try. Even if it never works, I'm going to keep trying." Stewart has probably done more than anyone to raise the public profile of "old friends" therapy, and he hopes that someday a drug will be developed from helminths that will be easier for people to take. As he speculates: "What if these things down-regulate the immune response? You could have a world where, at no risk, maybe you wouldn't have autism—you wouldn't have any autoimmune disease. Maybe I'm just too close to it, but it seems every time you turn over a rock, it still fits, every time something new comes up, it just fits this model."

It is tempting to get carried away by the hubris surrounding "old friends" therapy and react to Johnson's optimism by rushing off to declare a toast "To absent friends!" be they parasitic worms, friendly bacteria, or a whole host of microorganisms from the natural environment. Then raise our glasses full of probiotic drinking yogurt or whipworm eggs, quaff them down, and quickly replenish our "old friends" as soon as possible. But it might be wise to exercise a little caution at this stage and keep the celebrations on ice. Despite the promise, there are likely to be many hard miles to be traveled before the "old friends" hypothesis fully materializes as a reliable, effective, tried-and-tested evolutionary medicine of the future.

A FINE ROMANCE

HOW EVOLUTIONARY THEORY EXPLAINS INFERTILITY AND DISEASES OF PREGNANCY

> A fine romance, with no kisses
> A fine romance, my friend this is
> We should be like a couple of hot tomatoes
> But you're as cold as yesterday's mashed potatoes
> A fine romance, you won't nestle . . .

Let's face it, there is nothing in human life more likely to get us grasping for our rose-tinted glasses than the idea of a happy couple settling down together and beginning to make babies. The secret gasp of delight when the pregnancy kit indicator turns blue, the whispered sharing of the good news to a beloved partner, the wildfire family bush telegraph, a blooming pregnancy, a safe delivery, a pink or blue nursery, and a baby snuggling against its mother's breast: the romance of reproduction.

But try selling that romantic version of pregnancy to thousands of women every year who have their dream pregnancies turned into frightening nightmares ridden with raging high blood pressure, chronic gastric pain, kidney damage, and stressed babies thanks to the onset of a poorly understood disease of pregnancy called preeclampsia. When preeclampsia becomes severe, the only way to cure the mother is to deliver the baby, and many pregnant women have eventually lost babies to preeclampsia because they had to be delivered prematurely.

Our rose-colored reproduction romance also gets short shrift from Priya Taylor, whose attempts to start a family were traumatized by eight miscarriages in succession, six spontaneous miscarriages following IVF treatment, and the loss, early in pregnancy, of one twin before, finally, after the most turbulent pregnancy imaginable, she successfully gave birth to the other—baby Maia. Her courage and fortitude in the face of such appalling experiences almost defies description.

Surveying our crowded planet, with its 7 billion human inhabitants, you might be forgiven for having an inflated impression of human reproductive efficiency. In fact, we humans have surprisingly low fecundity. As Nick Macklon, professor of reproductive medicine at the University of Southampton, puts it: "When human reproduction is examined in terms of efficacy, it becomes clear that the growth in population has occurred in spite of, rather than as a result of, our reproductive performance." This was graphically highlighted, he says, when the number of births registered in England in 1970 was compared with the number of births that might have been expected given the estimated number of fertile ovulatory cycles exposed to coitus in the same year and population. It was a surprisingly dismal 22 percent and compares with up to 70 percent in cows, 60 percent in rabbits, and over 50 percent in dogs and many species of monkey. Translating those figures into normal sexual activity has suggested, by one estimate, that on average, although there are huge individual differences, a couple will need a hundred copulations to achieve a pregnancy, or a period of about seven to eight months of sustained sexual activity without contraception from the beginning of sexual relations. Part of the problem is the very high rate of miscarriage, the most common complication of pregnancy. About 30 percent of embryos are lost prior to implantation and a further 30 percent inside the first six weeks of gestation, most before the due date of the next menstruation. These pregnancies may be so short that a woman is oblivious of the fact that she has been, technically, pregnant. All she may be aware of is a heavy period. In addition, over 10 percent of pregnancies result in clinical miscarriage, mostly prior to twelve weeks gestation, and 1–2 percent of couples experience recurrent pregnancy loss, defined in the United States as failure of two or more consecutive pregnancies.

Even if a pregnancy progresses beyond twelve weeks, it can turn out to be a dangerous business. Gestational diabetes affects between 4 and 20 percent of pregnancies worldwide, while 10 percent of all pregnancies experience extremely high blood pressure, particularly in the third trimester.

This can damage the delicate glomerular structure of the kidneys, releasing protein into the blood. It can progress to HELLP syndrome, where the liver is damaged, and full-blown eclamptic fits involving brain seizures and convulsions. In the days before modern medical intervention, and in parts of the world today where medical services are poor, this can be fatal. Indeed, the emergency Cesarean section routinely performed today to deliver a baby and thus bring preeclampsia to an end in the mother originates in the days of the Roman Empire two thousand years ago as a last-ditch attempt to save the life of a baby whose mother was perceived to be dying from eclamptic convulsions. Even today, preeclampsia is the leading cause of maternal mortality worldwide—it accounts for up to 20 percent of all deaths.

Athena Byford had severe preeclampsia during her first pregnancy in 1998. She was twenty-six weeks pregnant at the time and remembers waking up one night with excruciating gastric pain and later, the following day, feeling very nauseous and suffering from a bad headache. These are all recognized symptoms of preeclampsia. It was Christmas, the doctor's office was closed, and so she had to soldier on to early January for an appointment. The nurse then took a blood sample and checked her blood pressure, then checked it again, before quietly slipping out to return with the doctor, who took her blood pressure once more. Covering up his concern, for fear of stoking her anxiety, he told her: "I'm going to write a little note and I'd like you to go home, get a bag packed, and pop up to the hospital. Nothing to worry about." At the hospital they read the note, looked at her, and said, "This can't be right." They told her to sit down, took her blood pressure again, and then, in Athena's words, "All hell broke loose. I was soon surrounded by nurses, midwives, and doctors, and I was being wheeled through the wards to the high-intensity care ward with strict instructions being given that I must not be left on my own under any circumstances. That's when I really started to worry." Within minutes she was hooked up to a number of machines and catheterized. A consultant arrived and told her "Your BP is sky-high—you're at risk of going into a convulsion. We don't want you going into an eclamptic fit, so we're going to give you an anti-convulsant—magnesium sulphate—and morphine." Slipping in and out of consciousness, she woke in the middle of the night to find a crowd of doctors and nurses around the monitors. Clearly unhappy about the way things were progressing, they told her the baby was now in distress and they wanted to take her straight to an operating room to deliver. Her baby girl duly arrived at twenty-eight weeks weighing just two pounds. "She was very, very small. I saw her briefly when she was born, because I was awake during the Cesarean, and they whisked her straight off to the neonatal unit."

Athena's blood pressure was still dangerously high, and so she wasn't allowed to visit her baby. Both the hospital staff and her family took photographs and a video to show her because in the middle of all this it was her birthday. The next day came bad news. Her baby had picked up an infection and had started to go downhill. She was on as much oxygen as they could possibly give her for fear of bursting her tiny lungs, and, finally, the doctors gently recommended she be taken off life support. "So they did that and they brought her up to me that evening in a Moses basket. They left her with me to hold until she did take her last breath. They offered to let me bathe her and to change the nappy after she had died, and they took a lot of photographs, which they said would help with the healing process—which it didn't—all I saw, to be honest, was a dead baby. I had to give the photos to a relative because I could not bear to look at them."

Evolutionary biologists, like women who have suffered from diseases of pregnancy, find the romantic view of pregnancy unappealing. They tend to strip away the emotional baggage of a woman's relationship with a loving partner, or a pregnant woman nurturing the baby inside her, and look beneath at the hard, rational logic of the genetic interests at stake in reproduction to explain why human females have such low fecundity; why there is so much apparent wastage of fertilized eggs and embryos; and why, for so many women, pregnancy is marred by life-threatening illness and potentially fatal fetal stress. Foremost among them is David Haig, from Harvard University, who has placed all these complications of pregnancy inside a theoretical framework called parent-offspring conflict.

On the outside, making babies seems a very cooperative effort, but in reality the genetic interests of mother, father, and fetus are not identical. Any fetus will, of course, inherit 50 percent of its genes from its mother, but it will also inherit 50 percent from its father. Throughout the animal kingdom, but especially so in humans, the female shoulders far more of the cost to bring any baby to term and nourish and care for it after birth than does the father. Male investment—in the shape of sperm—is literally microscopic. Furthermore, every baby a mother brings into the world will be genetically related to her, but they may have different fathers. It is therefore in the mother's best genetic interests to temper the investment she makes in any one baby in favor of spreading that investment out over any hypothetical number of babies she may incubate over her active reproductive life. Don't put all your eggs in one basket. It is in the selfish interest of paternal genes, in the shape of the fetus, to demand more from the mother than she is inclined to give. For a mother, the loss of one child, however upsetting, can be compensated for by any number of later pregnancies, by any number of partners, but for any fetus it is an existential matter—grow or die. This sets the scene for conflict.

Evolutionists would expect paternal genes, in the form of sperm and their genetic contribution to embryos, to favor mechanisms that promote the receptivity of the uterus and allow indiscriminate implantation of the embryo whether it is entirely viable or not. They would expect maternal genes to favor mechanisms that discriminate between good- and poor-quality embryos to reduce the chances of frittering away maternal investment on offspring that contain some genetic flaw or are otherwise incompatible. Once pregnancy has been established, they would expect paternal genes at work in the fetus and placenta to attempt to manipulate the mother to give up more food reserves than is in her long-term interest to do, and for maternal genes to resist that manipulation. Haig has likened this conflict of interest to a tug-of-war. Imagine two sides of brawny, sweaty men. If the sides are well matched, the flag at the center of the rope barely moves, despite all the straining and heaving. So it is with normal pregnancy—underneath the surface is the symmetrical pitting of opposing interests without which no pregnancy can be successful. The system would collapse if one side stopped pulling.

The great pioneering immunologist Sir Peter Medawar, working in the immediate postwar years and throughout the 1950s and '60s, did fundamental research to improve our understanding of how the immune system accepts or rejects skin grafts and transplanted organs. In turning his attention to pregnancy, he wondered how it could be that the immune system recognizes and routinely attacks the antigens (foreign proteins) present on grafted or transplanted tissue, and yet the maternal immune system seems willing to accept, tolerate, and host embryos and fetuses, despite the fact that they present foreign paternal antigens. Why doesn't the mother automatically reject the "semi-allogeneic" fetus? Medawar's answer was that the mother's immune system must somehow remain ignorant of the father's antigens. This could be either because there was physical separation of the fetus from the mother's immune system, or because the fetus was immunologically immature, or the mother became somehow incompetent to respond to fetal antigens.

Medawar's observations have stimulated much research on fetal tolerance in the past fifty years, and reproductive scientists are closing in on a comprehensive explanation for how it works. In doing so, they have finally rejected all of Medawar's suggestions because it has become obvious that there is no impregnable barrier between mother and fetus—there is leakage. Paternal antigens, shed in cellular debris from the fetus and the placenta, can be found in the mother's circulation throughout pregnancy and beyond. In fact, the maternal immune system is already well aware of paternal antigens even before an embryo attempts to implant itself in the uterine wall.

It has long been noticed that women are more likely to suffer preeclampsia if they get pregnant after a short period of cohabitation with a sexual partner than if they cohabit for more than six months before conceiving. Subsequent pregnancies with the same partner tend to be at lower risk of preeclampsia, but that risk climbs again if a woman changes partner between pregnancies or allows a lapse of a number of years between pregnancies. There is a greater risk of preeclampsia in a first pregnancy if the couple use condoms prior to trying to make a baby or if the frequency of lovemaking is relatively low. IVF pregnancies carry a higher risk of preeclampsia especially if carried out with donor sperm, and this risk is lowered if the couple concerned has unprotected sex frequently around the time of attempted IVF conception. There is even evidence that women who swallow their partner's semen during oral sex experience a lower risk of preeclampsia. All this suggests that components of semen and/or sperm can communicate with a woman's immune system, and that through repeated exposure to her partner's ejaculate, it learns to recognize her partner's antigens and develop a tolerance to them. From the introduction of semen into her reproductive tract, to the fertilization of her eggs, implantation of the resulting embryos, and the development of a fetus, paternal genes wage strategies to get themselves into the next generation, and maternal genes wage strategies to discriminate between partners and embryos and "decide" which embryos they will accept and how much they will invest in their nurture.

Semen is much more complicated than a mixture of sperm and nutrient seminal fluid. It is an extraordinary cocktail of active biochemicals. The psychologist Gordon Gallup, from the State University of New York at Albany, together with colleagues Rebecca Burch and Lori Petricone, has explored this emerging field of semen chemistry to reveal the active ingredients that manipulate a female's reproductive biology, and how she responds to them.

The vagina, says Gallup, is an ideal route for getting chemicals into a woman's bloodstream. Not only is it richly endowed with blood vessels, but blood from the vagina goes straight back to the heart via the iliac vein, bypassing the liver, which would normally break down imported products. This means that within an hour or two of insemination, you can measure elevated levels of semen chemicals in a woman's blood circulation. Many are designed to aid fertilization and the implantation of a newly fertilized embryo. In artificial insemination techniques that use "washed" semen, he explains, the likelihood of impregnation and subsequent fetal growth is reduced. Gallup also reports that researchers in one trial told half the women undergoing GIFT (gamete intrafallopian transfer) to abstain from lovemaking before and after the procedure and told the other half to make love close to the time of it. Fifteen of the eighteen women who had intercourse became pregnant whereas only five of eighteen abstainers did.

Semen also contains surprising amounts of hormones we normally associate with women. These include estrogen; follicle-stimulating hormone (FSH), which stimulates the recruitment, development, and maturation of follicles in the ovary; and luteinizing hormone (LH), which surges to trigger ovulation. Human semen also contains a range of cell-signaling molecules, or cytokines, including interleukins 1, 2, 4, 6, and 8, tumor necrosis factor-alpha, interferon-gamma, and granulocyte-macrophage colony-stimulating factor (GM-CSF), which all have immunosuppressant properties and are associated with making the uterus receptive to implantation. Gallup reports that semen contains levels of hormones that often exceed those found in non-pregnant women and even some cases of pregnancy. Chief among these is human chorionic gonadotropin (used as a sensitive indicator in medical tests for pregnancy), which maintains the corpus luteum in the ovary after ovulation, thus keeping progesterone levels high, which is vital for maintaining pregnancy.

Semen also contains thirteen types of prostaglandin and lipid messenger molecules that have been seen to reduce the activity of natural killer lymphocytes, one of the attack squadrons of the immune system. This range of cytokines and prostaglandins in semen, says Gallup, transfers to receptors on target cells in the uterus and cervix, where they affect gene expression so as to modify female reproductive tissues. The idea, he explains, is to increase sperm survival rates and fertilization, condition the female's immune responses to tolerate semen and the fertilized egg, and bring about changes to the uterine endometrium that will aid embryo development and implantation.

Sarah Robertson and her colleagues from the University of Adelaide have led research to work out exactly how these components of semen interact with the maternal immune system. Much reproductive biology research is conducted on mice, and Robertson's group wanted to know whether the way that seminal fluid elicits a mild inflammatory response in the murine female genital tract, activating immune reactions that enhanced the likelihood of conception and pregnancy, translated to humans. Medical ethics forbade invasive medical research on the uterus, and so they used the cervix as a proxy, arguing that immune reactions in the cervix would be mirrored in the uterus.

They selected a group of women who were all demonstrably fertile and divided them up into three groups. They asked them to abstain from sex for two days before a first biopsy, and to use condoms for the five days before that, ensuring complete absence of sperm in the reproductive tract. They then took tiny needle biopsies from the cervix at the time of ovulation and repeated the procedure two days later. In the intervening period, one group was allowed to have unprotected sex, the second group used condoms, and the third group abstained from sex altogether. They saw a wide number of immune events in cervical tissue that were unique to the group that had unprotected sex and, therefore, where the cervix had come into contact with semen. In this semen-exposed group, they saw a classical inflammatory reaction involving different populations of immune cells, changes in the activity of genes associated with inflammatory pathways, and changes in the active amounts of a number of pro-inflammatory cytokines. This did not happen in the abstaining group or among women using barrier contraception. A number of different types of white blood cell poured into the cervical epithelium, including macrophages, dendritic cells, neutrophils, and T lymphocytes—the memory cells of the immune system that specifically recognize foreign antigens.

The job of dendritic cells, as we discovered in the first chapter, is to process the antigens, or foreign proteins, on the surface of bacteria, virus-infected cells, and, in this case, sperm and semen, and then re-present them in such a way that they can be recognized by the effector T cells of the host's adaptive immune system. Depending on the antigen signal received, these T cells can be cytotoxic and will attack and destroy invading cells; or they can be pro-inflammatory, encouraging a hostile environment for future attempted implantation; or regulatory, providing a benign environment. It is the mild inflammatory reaction in the uterus, induced by the chemokine and cytokine chemical messenger molecules in seminal plasma, that recruits immune cells to the cervix or uterus, where this vital interaction can take place. The resulting immune memory means that when the woman's immune system meets those same paternal antigens on fertilized eggs and embryos, it will tolerate them and allow implantation to proceed. This is why extensive exposure to partner's sperm seems to give important protection against later pregnancy complications. Without this "tolerization," no pregnancy could be achieved.

Males are also found to vary considerably in the amounts of a cytokine called transforming growth factor-beta, or TGF-β, they contain in their semen. TGF-β is "immune deviating," meaning that it can skew a maternal immune response away from a hostile inflammatory reaction and toward a benign, tolerant environment where the population of regulatory T cells predominates. But this immune awareness of paternal antigens cuts both ways because it also allows females to discriminate between embryos for either quality or compatibility. For instance, one of the main ways that immune cells distinguish between "self" and "non-self" is via the major histocompatibility complex, or MHC. These are a variety of proteins (they are also called human leucocyte antigens—HLA—in us) produced by 160 MHC genes that are highly variable. So your MHC signature will be different from mine. In the world of organ transplantation, MHC-matching, or close affinity, is important to prevent rejection (hence the use of family members as kidney donors), but in the uterus, dissimilarity is the name of the game. There is evidence that if the mother's immune cells sense an MHC signature in semen that is very similar, they will reject cells bearing it. This is because it is the MHC molecules that actively present foreign antigens onto cell surfaces. Significant overlap between maternal and paternal MHC will narrow the range of antigens the resulting offspring can react with—which has clear implications for disease resistance. Furthermore, a closely matched MHC signature may betray the presence of shared potentially dangerous recessive genes that, in the resulting children, would become homozygous and therefore active—rendering all progeny susceptible to genetic disease. Not surprisingly, there is evidence that closely matching MHC between a woman and her partner results in increased rates of spontaneous miscarriage.

Let us imagine that an egg has been fertilized and begun to divide and nestle to the wall of the uterus as a prelude to implantation. Here is where the story of male-female conflict of interest gets positively Machiavellian. It has been known for some years that the first few cell-division cycles of a developing embryo or blastocyst derived from IVF are notoriously prone to chromosomal instability. This was thought to be one of the main limiting factors for IVF success and that the chromosomal instability somehow arises through the chemical hyperstimulation of the ovary to increase follicle and egg production. Because of this, the holy grail of infertility medicine has been to try to find methods that will result in the perfect embryo.

However, in 2009 Joris Vermeesch, from the Catholic University of Leuven, set out to develop a very sensitive screening test for chromosomal abnormalities in early embryos, and he took quite normal, naturally produced embryos from young women under the age of thirty-five who had no history of infertility. To his surprise, he noted a similar amount of "genetic chaos." He examined each of the cells, or blastomeres, that go to make up a three- to four-day embryo and found that over 90 percent of human embryos are genetically abnormal. About 50 percent had no normal diploid cells at all. Instabilities ranged from aneuploidy, where cells either had more or less than the normal complement of chromosomes, to uniparental disomy, where the cell received two copies of a chromosome from one parent and none from the other, to a whole mishmash of chromosomal deletions, duplications, fragmentations, and amplifications. You would think that an embryo so heavily compromised would be completely unviable. However, although the rate of pregnancy loss in humans through failure of implantation, spontaneous early miscarriage, and clinical miscarriage is extremely high at about 70 percent, it falls far short of the proportion, over 90 percent, of genetically chaotic embryos. So the number of perfectly healthy live births vastly exceeds the number of normal embryos.

This suggests several possible explanations. It may be that some chaotic embryos are able to correct themselves by the death of abnormal blastomeres, leaving only normal blastomeres to go on to form the fetus and placenta. It has been reported, for instance, that one frozen-thawed human embryo contained only one normal cell and yet went on to form a perfectly healthy baby. Or an embryo might have the ability to "self-correct" genetic mistakes. So, if a quantity of chaotic embryos manages to survive, why do they routinely go through a period of genetic instability in the first place? Jan Brosens, professor of reproductive medicine at the University of Warwick, and his colleague Nick Macklon in Southampton believe it is a strategy to make the embryo more invasive. The only other instance of such high levels of genetic instability comes from cancerous tumors, where it is invariably a prelude to aggressive cancer cell behavior including a transition to more motile cells, invasiveness, and metastasis, whereby cancer cells spread to secondary organs. The suggestion is that very early embryos can temporarily transition to the reproductive equivalent of malignant cancer. The finding of such high levels of "genetic chaos" in embryos is a genuine medical mystery that urgently warrants further investigation.

In the early embryo, some cells will go on to form the fetus while others will form the placenta, and humans have one of the most invasive placentas in the animal kingdom. This so-called hemochorial placentation, if successful, will force itself deep into the uterine wall and will eventually so profoundly alter maternal blood circulation that the mother will not be able to withhold nutrients from the fetus without starving herself to death. It is the ultimate parasite. Macklon points out that embryos are such inherently aggressive interlopers, they don't even need a uterus. Ectopic pregnancies occur when embryos manage to implant outside the normal confines of the uterine wall, usually in the Fallopian tubes, but occasionally in the cervix, ovaries, or even the abdominal cavity. Most of them are not viable, but they can be dangerous to the mother because their aggressive burrowing into blood vessels can cause extensive bleeding. There have been occasional miraculous cases where an ectopic pregnancy gave rise to a healthy baby. Perhaps the most famous is the case of Sage Dalton, who was born in 1999. She developed outside the womb because her placenta managed to burrow into a benign fibrous tumor in her mother's abdominal cavity that was richly endowed with blood vessels. Hydatidiform moles are another example of aggressive invasion. Here, on rare occasions, a sperm fertilizes an egg that has lost its maternal DNA. The sperm then reduplicates to form the diploid set of forty-six chromosomes—but they are all of paternal origin. It grows into a large, disorganized cellular mass in the uterus. It is, in effect, a placenta without a fetus.

Because of the relatively low metabolic cost involved in making sperm, males can afford to further their genetic interests by attempting to get as many embryos implanted as possible regardless of quality or eventual viability—a long-shot gambling game with big odds against it. The Trojan horse of genetically chaotic embryos is one example of this. Parent-offspring conflict theory predicts that females will have evolved effective countermeasures because the challenge for the mother is to prevent poor-quality embryos from gaining a foothold and wasting her precious time and energy, and to select among genetically abnormal embryos to discriminate those that have the potential to normalize. The metabolic cost to the mother, thanks to the evolution of the hemochorial placenta, is potentially enormous if she is unable to sift out unwelcome embryos. According to Brosens and his team of collaborators, this embryo selection is exactly what has happened thanks to the evolution of menstruation and the creation of a very narrow window in the menstrual cycle when effective implantation can take place. Earlier theories have tried to account for the proposed adaptive value of menstruation by suggesting it evolved to protect the female reproductive tract against sperm-borne pathogens, or that menstruation was less metabolically costly than constantly maintaining a thick uterine wall. But Brosen's work suggests that menstruation comes hand in hand with the evolution of a process called spontaneous decidualization, which allows human mothers to quality-control invading embryos.

Over one hundred animal species—including rodents, bears, deer, and kangaroos—experience diapause in reproduction, where an embryo can stick to the uterine wall and then arrest its development until it gets a "green light" from the uterus to begin implantation. For instance, in certain deer species, the embryo will arrive in the uterus in the autumn and implant in the spring. At this point the embryo initiates the maternal pregnancy response by inducing a process called decidualization. The decidua, or lining of the uterine wall, is so-called because it is shed with the placenta during delivery or, in humans, during menstruation. Where humans stand out—in company with Old World monkeys, the elephant shrew, and the fruit bat—is that they mount a pregnancy response whether an embryo is present or not. This is spontaneous decidualization, and it occurs between five and seven days after ovulation in every menstrual cycle. Brosens calls this the implantation window, or window of receptivity.

The uterus is supplied with blood through specialized blood vessels called spiral arteries, and the fibrous stromal cells that surround them undergo significant changes to differentiate into decidual cells and become secretory. Once this process starts, a continuous supply of progesterone is needed to keep it going. This is produced by the corpus luteum of the ovary. If no pregnancy occurs, progesterone levels fall and the woman will menstruate. However, if there is an implantation, progesterone levels remain high and the decidual cells in the uterine wall will migrate to encapsulate the conceptus and recruit immune cells to the site so that interrogation of the conceptus can begin. It is the biological equivalent of the inquisition of a prisoner in a police cell.

Several features suggest that the adaptive value of menstruation is that it preconditions the uterus to temporarily go through these decidual changes and prepare itself to incorporate and then interrogate embryos. Baby girls frequently experience menstruation during the first few days of life, after which the uterus becomes quiescent until menarche. At menarche, it is usual for young women to menstruate for up to two years before they begin to ovulate. Brosens and his colleagues believe it is no accident that menstruation so precedes pregnancy: it is the intense bleeding and inflammation of menstruation that does the preconditioning of the uterus. Repeated, regular menstruation also makes sure that the transient inflammatory reaction to an attempted implantation is attenuated, sufficient to stimulate the necessary morphological and immune changes in the uterus without being severe enough to prejudice the embryo. This is the mild inflammation also documented by Sarah Robertson and may explain why severe preeclampsia is a disease of first pregnancy and disproportionately affects young women. You can trace it back, explains Brosens, to insufficient menstrual preconditioning of the uterus.

To this day, says Brosens, medical textbooks talk of active, invading embryos and a passive endometrium, or uterine wall. This medical metaphor has a long history. David Haig has documented turns of phrase commonly used by reproductive scientists in the first two decades of the twentieth century that are clearly infected with the zeitgeist of impending war. For instance, Ernst Gräfenberg, writing in 1910, described the ovum as "*Ein frecher eindringling*"—a cheeky intruder eating its way deep into the uterine wall, while Oskar Polano described supposed antagonism between maternal and fetal tissue by describing the latter as "establishing outposts in enemy territory." "On the eve of the conflagration that was to engulf Europe," writes Haig, "Johnstone used eerily prescient language: 'The border zone is not a sharp line, for it is in truth the fighting line where the conflict between the maternal cells and the invading embryo takes place, and it is strewn with such of the dead on both sides as have not already been carried off the field.'"

In truth, the cells of the decidua play a very active role, but it is in selection, not trench warfare. The abnormal embryo produces clear chemical and immunological signals of its incompetence, but the mother can only recognize those signals by going through the decidualization process. Thus, in humans you not only have a window of receptivity when the embryo can implant, but it is also a window of embryo recognition and selection. And it is the same process, says Brosens, that will ultimately control how the placenta is formed. So if you have patients who have poor decidualization, you will have poor embryo recognition, a compromised placenta (even if the embryo is competent), and either early embryo loss through spontaneous abortion or preeclampsia later in the pregnancy.

In his day job, Brosens sees many women with fertility problems. They either appear infertile or super-fertile but unable to hang on to the baby. He remembers examining a Scottish woman some years ago who had had eight miscarriages. Since it was widely assumed at the time that infertility and miscarriage were strongly linked, he asked her how long it took for her to get pregnant, assuming he would hear a tale of frustration. He was shocked when she replied, "Always one month!" This woman had no fertility problems whatsoever; she got pregnant like clockwork at every cycle in which she tried, but every time the fetus was lost in early miscarriage.

In 1999 Allen J. Wilcox, the scientist who had first revealed the very high rate, at over 30 percent, of normal early pregnancy loss, published a paper in the *New England Journal of Medicine* that looked at the fate of implantation dependent on how many days after ovulation it occurred. He recruited a very substantial number of women who were trying to get pregnant and persuaded them to fill their refrigerators with urine samples taken every day until a successful pregnancy was achieved. His team analyzed the urine to discover when the surge of luteinizing hormone occurred, which gave them the likely day of ovulation. They then measured levels of human chorionic gonadotrophin (HCG), which is a sensitive indicator of implantation. Most women implanted within a narrow window about six or seven days after ovulation, but there were a number of women who implanted much later—even eight to eleven days beyond. Wilcox showed there was an exponential increase in miscarriage with late implantation. Brosens believes this is because late implantation misses the crucial window of receptivity in which decidual changes in the uterine wall prepare the mother to envelop and examine implanting embryos like a stern schoolmistress. Late-implanted embryos fail to turn up for this time-limited examination, are deemed to have failed the exam whether they are competent or not, and expelled. Women who allow this late implantation have developed a fault in an evolved female mechanism for embryo quality control that then reveals itself as reproductive illness—super-fertility allied to recurrent miscarriage—as Priya Taylor found out to her cost.

Priya wanted to start a family immediately after she got married in 2003, and she got pregnant on her honeymoon. However, between ten and twenty weeks later, she started intermittent spotting and at about twenty-two weeks, she recalls: "I woke up and my stomach just seemed smaller. My husband said 'Something's not right.' I felt like I'd wet the bed. My waters had broken." Her highly premature baby was eventually delivered at twenty-five weeks. "It's a sad story really. Alexander cried when he was born. They didn't warn me they were going to take him away from me. They were hiding behind a curtain—I didn't even know they were there. They put him into Intensive Care. He was tiny—one pound—and needed help breathing. He made it through the first two days—and then he died."

Undaunted, Priya got pregnant again within two months of losing Alexander, but at ten weeks the doctors could detect no fetal heartbeat. She had a dilation and curettage (D and C) operation to remove fetal and placental tissue and by March was pregnant again. Another failure was quickly followed by six more, all of which lasted between four and ten weeks. "Waving me off to successive D and Cs was becoming very hard for my husband, and I was watching all my best friends having children. It was really terrible. I'm quite an emotional person—I get upset very easily—but I also can't bear to be felt sorry for. And I can't bear to think there's something I can't control—or do."

Priya was still trying to get pregnant naturally but nothing was happening. So, two months later, they started their first round of IVF—it was to be the first of six—and five of them led to short-lived pregnancies. She even had a natural pregnancy in between that lasted about seven weeks. At that point Jan Brosens took over her treatment for a while. Laboratory analysis of tissue from her fifth fetus showed that it had a chromosomal abnormality and was never likely to survive, and although she had high hopes for number six, her next miscarriage was a horrendous experience: a canceled Caribbean holiday at the last minute, a D and C that went wrong, and a severe uterine infection that forced her, bleeding, into the hospital for a week. Her husband, Matt, normally a typically stoic male, cried for the first time throughout this very trying week and said: "That is it. No more. Nothing you say to me is going to convince me to do this again." By now she was almost broken and at the end of her tether, but at their wedding anniversary eight weeks later, she pleaded, "Please, can we just go one more time, and then I promise that's it." It took Matt about a month to agree.

Brosens believes Priya is at the extreme end of the super-fertility spectrum, a uterine disorder typified by prolonged receptivity, impaired decidualization, and lack of embryo selection. Her evolved mechanism for embryonic quality control had developed a fault, and the clinical consequences for Priya were rapid conceptions but high incidence of early pregnancy loss.

Priya's next—and last—IVF cycle produced an astonishing twenty eggs of which fourteen fertilized. Two day-six embryos were introduced. She shortly got a phone call saying her pregnancy test was positive and the readings were really high, suggesting a successful implantation. However, three days later, the levels started to drop. She hung in there and saw the heartbeat at six weeks but still didn't believe the pregnancy had survived. However, Brosens, who was looking after her day-to-day care, scanned her every week between six and twelve weeks just to prove to her that the baby was growing. In fact, she was nursing two fetuses, one of which aborted at about seven weeks, but baby Maia clung on. At sixteen weeks they stitched the neck of her cervix to stop the baby from coming out, but at eighteen weeks she began to bleed. "Again—middle of the night—my bed is full of blood. We get up and I say to Matt, 'Just pack a bag—I'm going to be in for a while and the baby will not be coming home with us.'" But when they got to the hospital, the baby still had a heartbeat. Brosens scanned her the next day, and he could see quite a big blood clot near the placenta. At twenty-two weeks her baby stopped growing; the placenta was highly abnormal with insufficient connections to the maternal blood circulation inside the womb. It was also covering the whole of her cervix, meaning a natural birth was out of the question. Still the pregnancy tottered on until finally, at thirty-five weeks, her doctors had seen enough of low growth rates and poor blood flow through the placenta. Baby Maia was finally born by Cesarean section at a healthy four pounds five ounces. After five days the medical team agreed to discharge her: "At which point they said to me: 'Is everything ready at home?' I hadn't got a thing—not a single thing! They said, 'Why not?' and I had to tell them that I had no idea I would be taking a baby home."

The cellular and immune changes in the uterine wall that determine either a successful or unsuccessful implantation are extremely complicated and the subject of widespread ongoing research with no little scientific disagreement. And, as yet, our knowledge of them is heavily dependent upon mouse experiments that we do not always know how to extrapolate to humans. Nevertheless, it is becoming obvious that impaired decidualization, involving abnormalities in the way that stromal fibroblast cells differentiate into decidual cells, not only affects a woman's ability to mount a receptive but inquisitorial response to an invading embryo, but lays the foundations for further diseases of pregnancy like early miscarriage and pre-eclampsia.

Brosens and colleagues Madhuri Salker, Siobhan Quenby, Gijs Tecklenburg, and others have looked at the process of decidualization in minute detail. The action of a cytokine, interleukin-33 (IL-33), one of a family of pro-inflammatory cytokines, seems crucial. Progesterone is vital for the differentiation of stromal cells to decidual cells. As they begin to differentiate, the cells secrete IL-33 and bind it to a receptor molecule STL2, which allows it to induce a cocktail of chemokines, cytokines, C reactive protein, and other inflammatory factors in the uterine wall. This is the transient inflammation that is a vital prelude to implantation because it switches on a number of uterine receptivity genes. The inflammation is self-limiting, however, because, as they achieve full decidualization under the influence of further progesterone, the cells activate a feedback loop and begin production of a decoy receptor for IL-33 called sST2. This effectively silences IL-33, and the inflammatory reaction comes to an end. When the scientists compared gene activity levels for STL2 in stromal cell cultures obtained from normal women versus women suffering from recurrent pregnancy loss, the STL2 levels rose initially in both, but in normal women tailed off after two days of decidualization. In contrast, they remained high for eight post-decidual days in recurrent pregnancy loss patients, and levels of the decoy, sST2, were significantly lower. Although these women were able to implant embryos in this prolonged window, the cumulative inflammatory effects of STL2, and the inflammatory cytokines it initiates, had gone on longer than they should have and produced a very hostile milieu for those embryos because the decidual tissue was now damaged and bleeding and populated with immune attack cells.

As they transform from stromal cells, the changes in the type of chemical signaling molecules—cytokines—that the decidual cells produce enables them to act as gatekeepers of this crucial uterine interface between mother and embryo. By secreting interleukins 11 and 15, they encourage the recruitment and differentiation of a specialized type of white blood cell called a natural killer, or NK, cell. Normally, in peripheral circulation, these cells, as their name implies, are extremely toxic and will destroy virus-infected or tumor cells. However, uterine NK cells (uNK cells) function differently and, instead of producing cell toxins, produce a range of cytokines that seem vital for pregnancy to continue. The NK cell population rises to reach 70 percent of all white blood cells in the outer uterine wall, or decidua, and they help to form the placenta by encouraging the growth of new blood vessels and the invasion of maternal tissue by the embryo. At the same time, the decidual cells protect the fetus—allogeneic by virtue of the paternal genes it carries—from being attacked by cytotoxic T cells. They do this by shutting down their own key cytokine genes that would otherwise produce the chemical messengers that guide the T cells to their fetal target.

An embryo that has survived maternal interrogation now moves on to invade the uterine wall more intimately and deeply to form the placenta. As the cells in the embryo divide, they separate to form an outer layer of cells, called the trophectoderm, from which the invasive trophoblast cells develop, and an inner mass of cells that will form the fetus. As it pushes ever deeper into the uterine wall, the trophoblast produces tree-like branching structures, rich in blood vessels, called the chorionic villi, which anchor the fetus. The trophoblast, now called extravillous trophoblast, forges ahead and creates open endings to the spiral arteries in the uterine wall such that they drain, unfettered, into a large lumen called the intervillous space, at the boundary between maternal and fetal tissue, where maternal blood freely bathes the chorionic villi. A layer of cells called the syncytiotrophoblast lines the villi and the intervillous space. Now the spiral arteries undergo drastic remodeling. The smooth muscle and the elastic fibers of the artery wall start to break down and become replaced by fetal trophoblast cells. This dramatically transforms the arteries from high-pressure, muscular structures into low-resistance baggy vessels, the biological equivalent of worn-out panty hose.

By twenty weeks of normal gestation, an extraordinary structure has matured—the placenta. It has, at birth, a vast surface area of eleven square meters. That makes it almost as efficient as the lungs at gas exchange. Furthermore, the remodeling of the mother's spiral arteries means that it is impossible for her to restrict blood flow to the placenta by constricting them. Another important consequence of this deep placentation is that the placenta can release substances directly into the maternal bloodstream to manipulate her metabolism, while the mother is at a disadvantage in this regard because maternal substances must first cross the syncytiotrophoblast and then the fetal epithelium before they can gain access to the fetus.

The uterine wall, fetus, and placenta constitute an immune privileged zone where different rules apply. Instead of attacking invading cells that signal that they are "non-self," the uterus has evolved the ability to inspect and select the invaders before plotting either a hostile or benign response.

Placentation first evolved in mammals about 120 million years ago, and we humans share our particularly complex and deep hemochorial placentation only with the great apes and a few other species. Hemochorial placentation has allowed mammals like us to gestate their young for a long period of time, transferring a huge amount of food and oxygen to them in the process, but it has required very deep and dramatic levels of uterine invasion by the embryo's trophoblast, very significant amounts of maternal investment, and an immune system that, locally, suppresses its normal mode of action. Both the innate and adaptive immune systems are involved in the form of specialized populations of NK cells and T cells, respectively, which behave very differently to their relatives in peripheral circulation. For placentation to arise, a substantial amount of evolution in mammalian genomes was required, some of which was decidedly fortuitous.

We first encountered regulatory T cells (Tregs) in the previous chapter, "Absent Friends," where we discovered that they protect against allergic and autoimmune diseases by inhibiting the overproduction of effector T cells. They are also vital for immune tolerance to the fetus. They were originally called suppressor cells because early experiments on tissue grafting suggested that they could prevent rejection, but they fell out of favor thanks to hyperbole over their supposed widespread effects and technical limitations over their assay. It was the discovery of a vital marker gene in T cells, a transcription factor called FOXP3, that put them back on the map. Mice and humans that lack FOXP3 have no Tregs and suffer calamitously from autoimmune disease.

Just prior to ovulation, there is a large increase in regulatory T cells in the peripheral circulation. This may be driven by estrogen and progesterone levels, and may help to explain why pregnant women with rheumatoid arthritis, an autoimmune disease, often experience a remission in their symptoms during pregnancy. Tamara Tilberg's group at Harvard University has shown quite clearly that T cells can recognize specific variants of a group of human leukocyte antigen (HLA) molecules called HLA-C, the only group of HLA antigens that are polymorphic in that they can exist in 1,600 potential versions. This is an example of the histocompatibility mechanism described earlier. These HLA-C variants are displayed by embryonic and fetal tissue, and it is this ability to discriminate among them that allows the T cells to either react in a hostile or benign fashion depending upon what they reveal about the compatibility of paternal genes. In HLA-C mismatched pregnancies (where the paternal HLA-C molecules differ significantly from maternal

HLA-C), both cytokine-releasing T cells and regulatory T cell populations are increased. It is the presence of the regulatory cells that holds the potentially hostile action of the effector T cells in check, and a number of studies have shown that a reduced regulatory T cell population is associated with both recurrent spontaneous abortion and preeclampsia.

Robert Samstein, from the Memorial Sloan Kettering Cancer Center in New York, has produced a fascinating evolutionary dimension to the contribution that regulatory T cells (Tregs) make to immune tolerance of the fetus. Most Tregs are produced in the thymus gland in the neck, hence the "T." However, it has been discovered that a separate population of Treg cells can be produced from naive T cells in the peripheral blood circulation. It is only this Treg population that is involved in fetal tolerance, and Samstein believes it evolved specifically to reduce the conflict between mother and fetus that inevitably arose when female placental mammals were exposed to paternal antigens. Samstein has shown that the differentiation of these peripheral regulatory T cells requires the presence of the gene FOXP3 and a crucial non-coding genetic element called CNS1, which enhances its action. CNS1 is not required for the maturation of thymic T cells. Samstein looked for CNS1 in a wide range of species in the animal kingdom only to find that it emerged abruptly in the evolution of the placental mammals. It turned out to be a "jumping gene" (scientists call it a transposon) of the type first identified by Barbara McClintock in the 1950s. CNS1 had "upped sticks" from elsewhere in the genome and jumped onto another chromosome, where it landed just downstream of the FOXP3 gene in such a way that it could be co-opted by evolution as a gene enhancer. Samstein has reported a series of experiments on female mice showing that CNS1-deficient mice recruited far fewer regulatory T cells to the decidua. CNS1-deficient dams, even when they had been mated with MHC-mismatched males, showed early necrosis of the spiral arteries, inflammation and swelling, and resorption of fetuses.

This new model for the evolution of specialized regulatory T cell enforcers of fetal tolerance makes sense because it explains the observation we discussed earlier, that women who conceive quickly with a new partner are more prone to preeclampsia. It could be because they have not had enough time to develop a tolerance to the specific pattern of HLA-C molecules first presented to them on their partner's sperm. It could also explain why preeclampsia risk rises if there is a long time lapse between pregnancies—it could be because immune memory will have waned. It also helps to explain the importance of the very high levels of the cytokine TGF-β in human semen, reported by Sarah Robertson. TGF-β is also vital for the differentiation of the regulatory T cells that eventually become operational in the uterus. If regulatory T cells are absent in the uterine wall when the placenta is developing, there is nothing to suppress an immune reaction against the fetus.

As the fetus starts to develop, the tug-of-war between mother and fetus begins in earnest because, according to David Haig, it is now in the mother's interest to restrain the unbridled growth and greed of the fetus, and in the fetus's interest to take all the nutrients from its mother that it can get. Since the fetus carries two copies of all genes, one derived from its mother and the other from its father, it has representatives from both parents' genomes in its DNA. Evolution has found a resolution to this maternal-paternal conflict in mammals in the phenomenon of gene imprinting, which is the silencing of a gene's activity by methyl chemical molecules that attach to the DNA and inactivate it. If a gene is maternally imprinted, it means that the copy of the gene in the fetus that has been derived from its mother is silent, but the copy from the father is active. If paternally imprinted, it is the gene from the father than is rendered inactive. About 150 imprinted mammalian genes have now been found (many more will soon be identified), and many of them are involved in the placenta and fetus. As you might expect, they often act in opposition to each other—this is Haig's tug-of-war again. Scientists have done a number of experiments where they have knocked out either the maternal or paternal copy of these pairs of imprinted genes to see what happens when something goes wrong with the symmetry of imprinting. They are deliberately interfering with the symmetry of Haig's tug-of-war to see what happens when one side of the tug-of-war stops pulling.

For instance, one of the earliest pairs of imprinted genes ever found was insulin-like growth factor 2 (IGF2). This promotes fetal growth because, normally, only the father's copy is active; the copy from the mother is silenced. When scientists knocked out the paternal copy of IGF2 in mice, they tilted the balance back in favor of the mother's interests and the resulting fetuses were 40 percent lighter than normal. The mother normally counters IGF2 with an opposing gene called IGF2R that regulates fetal growth because the paternal copy is silenced. When the maternal copy of that gene was knocked out, the balance was tilted back in favor of the father: there was a 35 percent increase in the production of placental hormones, and the resulting mice were 125 percent normal weight.

Scientists from the University of Bath have recently extended this story of antagonistic genetic control over fetal growth to two more genes on opposite sides of the tug-of-war: Dlk1, which is active in the father but silenced in the mother; and Grb10, which is silenced in the father but active in the mother. Grb10 knock-out mice were up to 40 percent larger than their normal littermates and had high levels of fat deposition, while, conversely, Dlk1 knock-out mouse pups were 20 percent lighter. Both genes operate through the same genetic pathway, and so their antagonistic effects result in balanced, normal growth.

PHLDA2 is a maternally expressed gene that restricts placental growth, which explains why its overactivity has been associated with IUGR (intrauterine growth restriction) babies. One study also links its overactivity with increased rates of miscarriage and stillbirth, which may be because it interferes with the ability of the placenta to remodel the spiral arteries of the mother. The effects of PHLDA2 are countered by a gene called PEG10, which is paternally expressed and very active in the placenta. It has very low activity early in pregnancy but increases its expression at ten to twelve weeks of gestation and maintains that high level of output right up to term.

CDKN1C is maternally expressed, and if the maternal copy is disabled, it results in overgrowth of the placenta. Valaria Romanelli has looked at a number of women where mutations had disabled the CDKN1C gene. They eventually gave birth after very severe episodes of HELLP syndrome, and their babies were overweight and suffered from Beckwith-Wiedemann syndrome, which produces babies with large limbs and bodies, together with other defects. This suggested that the balance had been shifted in favor of paternal genetic interests, causing abnormal fetal growth and nutritional demand.

Two other linked disorders of imprinted genes show how balance is crucial for appropriate fetal development. Angelman syndrome is caused by mutations that inactivate maternally derived genes from chromosome 15 that are normally active while the paternal copy is silenced. Children born with Angelman suffer from severe sleep disturbance and long periods of uncoordinated suckling. Their angelic appearance and happy demeanor seem almost calculated to betray the existence of paternal genes manipulating maternal attention. Prader-Willi syndrome, in contrast, is caused by the inactivation of genes on chromosome 15 that are normally active if they have been derived from the father but silenced in the mother. These children are difficult to wake, physically inactive, and have very poor suckling reflexes. However, by their second year, when they are usually weaned, they develop voracious appetites and often become obese. In this case it is the mother's interests that have become favored because the syndrome results in poor suckling at a time when her food reserves would normally be enthusiastically consumed, while the baby's insatiable food consumption later on does not come directly from her own food reserves via breast-feeding.

David Haig has recently extended the above line of thinking to a phenomenon that any harassed mother knows only too well—an infant waking frequently in the night and demanding to be suckled. Since Angelman syndrome, where babies have highly interrupted sleep and long periods of suckling, is underpinned by paternal genes that have been released from imprinting by the inactivation of their maternal counterparts, he argues, it may be that these paternal genes in normal demanding babies constitute an adaptation in infants to extend their mother's period of effective lactation. This would suspend her return to normal ovulatory cycles and thus delay the birth of a second sibling—thereby reducing the chance of sibling rivalry for a mother's nutrition, care, and attention. Darwinian theory has become almost cloaked in Machiavelli!

By approximately twenty weeks of pregnancy, the remodeling of the mother's spiral arteries is complete. From now until she gives birth, in a normal pregnancy, the mother will increase her heart rate and the number of red blood cells in her circulation as her metabolism adapts to providing for her baby as well as herself. She can also expect some level of high blood pressure as even normal placentas release cellular debris into the maternal system that triggers mild inflammation in the blood vessels. Early-onset preeclampsia, which normally sets in about this time and is typified by inadequate placentation, is simply an enlargement of this inflammatory process, according to Ian Sargent and Chris Redman, from the Nuffield Department of Obstetrics and Gynaecology in Oxford. Some researchers believe that inadequate remodeling of the spiral arteries reduces blood flow to the placenta and induces hypoxia, but Sargent and Redman believe that it is not flow volume per se but intermittency in flow through narrow arteries that have resisted remodeling that is the important factor. This intermittency, they think, leads to transient ischemia in the placenta. When blood flow is restored, it suffers from reperfusion injury in exactly the same way that the heart is further damaged after a heart attack when blood flow is restored through the reopening of blocked coronary arteries. The sudden inrush of blood, oxygen, and nutrients, instead of being benign, causes inflammation and oxidative stress through the release of free radicals. The damaged placenta releases inflammatory factors and debris from damaged and dead cells that quickly produce a systemic inflammatory reaction in the mother's arteries, damaging their endothelial lining and raising her blood pressure.

Haig's theory predicts that if a fetus and placenta are compromised in this way, they should take the fight back to the mother by using very specific biochemical weapons to restore an adequate blood supply. He reasoned that the placenta should begin by increasing its mother's cardiac output, but if that proved insufficient, it should then try to raise the resistance, and hence the blood pressure, in its mother's peripheral circulation, which would have the effect of diverting blood into all her core organs, including the uterus and placenta. His theory has been vindicated by Ananth Karumanchi, a kidney specialist at Harvard Medical School.

Around 2000 Karumanchi began to see a number of women who were suffering from high blood pressure in pregnancy together with kidney failure. Intrigued by the apparent lack of good research and consensus on the causes of preeclampsia, he began a research program, using discarded afterbirths, to look at which genes increased their activity in diseased placentas where it was known that the protein the gene coded for could enter the mother's circulation. A protein called soluble fms-like tyrosine kinase 1 (sFlt1) figured very prominently in his results, and when he tested bloods from women suffering from severe preeclampsia, he found five times more sFlt1 in their blood than in women in normal pregnancy. Furthermore, when he administered sFlt1 to rats, they developed a range of preeclamptic symptoms. He published his results in 2003 and soon afterward found Haig knocking on his door. Haig was delighted to discover a real-life example of his theory, and Karumanchi has been a great Haig fan ever since because Haig has provided him with a theoretical framework that allows him to explain—short of outright pathology—why it could be in a placenta's interest to wreak havoc on its host's circulatory system.

The growth of new blood vessels and the day-to-day maintenance of the endothelial lining of existing blood vessels are governed by a protein called VEGF. A healthy arterial wall leads to normal blood pressure. But sFlt1 is a VEGF agonist; it binds to it and inactivates it. This drives up blood pressure in the mother's peripheral circulation, thus diverting more blood to the placenta, just as Haig prescribed. But VEGF is also important for maintaining the endothelium in the fine tubules, or glomeruli, in the kidneys, which filter waste products out of the blood, and it is also active in the liver and the brain. This explains why preeclampsia damages the kidneys and causes proteinuria. SFlt1 does not act alone. The stressed placenta also releases soluble endoglin, another factor that causes high blood pressure. This may act in concert with sFlt1 and is associated with the extremely severe form of preeclampsia, HELLP syndrome, where pregnant women develop headaches, severe heartburn, and raised liver enzymes.

When Haig was first formulating his hypothesis of parent-offspring conflict, back in the 1990s, he couched it in terms of the battle over glucose between the mother and the fetus. At least 10 percent of women will experience gestational diabetes, particularly during the third trimester of pregnancy. This inevitably goes away as soon as the baby is delivered. The blood-sugar levels rise because the cells of the body are becoming resistant to the effects of insulin—their insulin receptors are becoming less efficient. In response, the mother produces higher and higher levels of insulin to try to stabilize her blood sugar. This puzzled Haig. Surely, since insulin resistance and insulin production tend to cancel each other out, he thought, homeostasis of blood sugar could more economically be arrived at with less insulin resistance and lower insulin production. But this fails to take the placenta into account. If unopposed, it will always have the capacity to remove more glucose from maternal blood than is in the mother's best interests to give. The mother, via insulin production, therefore tries to limit her blood glucose.

After every meal, said Haig, the mother and fetus "squabble" over the share of glucose each will receive from it. The longer the mother takes to reduce her blood sugar, the greater is the share of it that will be taken by the fetus. In order to counteract the effects of maternal insulin in reducing glucose supply, you might expect the fetus to come up with a device to limit the effectiveness of insulin, and it seems this is exactly what happens. The placenta releases human placental lactogen, which interferes with the insulin receptors on its mother's body cells, blocking the action of insulin and causing blood sugar to rise. It also stimulates the production of a number of pro-inflammatory cytokines that also interfere with insulin action and promote insulin resistance and hyperglycemia. The mother produces more insulin as a countermeasure but, eventually, cannot keep pace, and gestational diabetes results.

How long should a normal pregnancy last, and why does it eventually come to an end? What actually brings childbirth about? Human gestation lasts, on average, about forty weeks, though there is substantial variation through premature delivery. Until modern obstetric practice intervened, some babies would go on to forty-two or forty-three weeks, though the heavy demands of such laggardly babies would invariably impose severe preeclampsia on their mothers. In 1995 Roger Smith and colleagues, from the John Hunter Hospital in Australia, proposed the existence of a "placental clock," active from the early stages of pregnancy, which determined the length of gestation and the timing of childbirth. Their clock depends on the action of a hormone called corticotropin-releasing hormone (CRH), which is produced in the placenta and released into the maternal circulation from about twenty weeks onward and which rises exponentially during the final trimester.

Throughout pregnancy, as CRH pours into the maternal circulation, a binding protein that is produced in the mother's liver immediately inactivates it. So, although the circulating concentration of CRH rises, it is in its inactive form. Eventually, about three weeks before parturition, the placental deluge of CRH overwhelms the production of binding protein and thus causes the exponential rise in active CRH. When the Australian scientists measured blood plasma CRH levels in pregnant women, they discovered that those women who subsequently delivered prematurely, at an average of thirty-four weeks, had much higher levels of circulating CRH than women who delivered normally at about forty weeks. Conversely, women who went beyond normal term, to an average of forty-two weeks, had much lower levels of CRH. The placental clock worked, they said, by the unmasking of CRH such that those mothers where CRH was unmasked from the binding protein earlier went into labor earlier and vice versa. CRH levels early in pregnancy, they concluded, were a reliable indicator of gestation length and an early warning sign for women who would likely experience premature labor.

Smith and his colleagues proposed that the high levels of active CRH late in pregnancy actually brought about childbirth because there are receptors for CRH present in fetal membranes and CRH stimulates the production of prostaglandins and oxytocin that produce the muscular contractions in the womb necessary to dispel the baby. This may be synchronized with the baby's development, they reasoned, because CRH also stimulates fetal adrenal hormones involved in organ maturation. The model they produced is essentially one of collaboration between the mother and the baby over the appropriate timing of birth.

In a further twist to this story, it is known that CRH is produced in all of us—including pregnant mothers—as a vital component of the stress response. It is produced in the hypothalamus in the brain and stimulates the pituitary gland to produce adrenocorticotropic hormone (ACTH). This travels to the adrenal glands, causing them to release corticosteroids, especially cortisol (sometimes known as hydrocortisone). This is the so-called HPA (hypothalamic-pituitary-adrenal) axis. As pregnancy progresses, the mother's HPA-mediated production of CRH, ACTH, and cortisol becomes dampened. The conventional view is that this is a maternal effort to protect the fetus, which was thought to be highly sensitive to cortisol.

However, a number of features of Smith's placental clock model puzzled Steve Gangestad, head of the evolution and development program at the University of New Mexico. He applied Haig-think to the situation. Why, if the mother is throttling back on her hypothalamic CRH production to reduce cortisol in order to protect the fetus, does the fetus shoot itself in the foot by pouring placental CRH into its mother's circulation? Furthermore, if the ultimate targets of CRH are the receptors in the placenta and its membranes, to jolt childbirth into action, why does the placenta not simply act locally, rather than deluging its mother with CRH throughout pregnancy? It suggested, to Gangestad, an alternative and more logical explanation.

Cortisol acts on the liver to increase blood-sugar levels to provide fast energy for the "fight or flight" stress response. The reason the placenta releases so much CRH, he argues, may well be yet another attempt by the fetus to lever more glucose out of its mother, since CRH equates with raised cortisol. The mother fights back by sequestering away as much CRH as she can by binding it in a protein complex and by decreasing her own production of CRH and cortisol. In support of Gangestad's reinterpretation of the placental clock hypothesis, intrauterine growth restriction of the fetus is associated with increased levels of CRH, as is preeclampsia. This is because the compromised fetus/placenta is fighting back with CRH. Resistance to blood flow to the placenta (as would occur thanks to poor spiral artery remodeling) is also associated with high levels of CRH while cortisol also causes constriction of maternal blood vessels by increasing their sensitivity to epinephrine and norepinephrine. This, again, would increase resistance in peripheral maternal circulation and would tend to shunt blood toward the compromised placenta.

No one disputes the fact that high levels of CRH in the second trimester are associated with pre-term birth, but Gangestad thinks that this may be because, at some point, the increasing demands of the fetus for nutrients exceeds the rate at which they can be transported across the placenta. At this critical juncture, he says, the fetus begins to mobilize its own fat reserves—

it is starving. It has reached a metabolic cross-over point at which it is now better to be born and to begin extracting further resources from its mother via breast-feeding, than to remain in the womb trying to bludgeon more nutrients out of its mother's circulation. In this interpretation, the fetus produces large amounts of CRH to squeeze more glucose out of its mother, but, eventually, if this proves insufficient, the persistent high CRH produces enough cortisol to initiate birth.

This idea chimes well with the "energetics and growth" hypothesis for gestation length and timing of parturition put forward by Peter Ellison and Holly Dunsworth. The timing of childbirth, they argue, has not evolved so much under the constraints of the size of the female human pelvis and what can be pushed through it—the so-called obstetric hypothesis—but the balance between maternal and fetal metabolism. In their model, birth occurs due to metabolic stress, which sets in when the mother can no longer meet fetal demands. Although many imprinted genes have been shown to be active in the placenta, there is evidence to suggest that many are also involved in brain activity. It could be that gene imprinting after birth represents a new battle in the campaign between the mother and her baby, this time over maternal care and further supply of nutrients via the breast. This would involve mechanisms of attachment to the baby, attractiveness of the baby to the mother, and stimulation of lactation.

This evolutionary view of pregnancy, gestation, and childbirth seems harsh and competitive, with talk of conflict of interests, and the description of the maternal-fetal interface in the womb, as David Haig puts it, being either fertile soil or no-man's-land. It is often confused with conscious human decision-making, in which case prejudicing a baby would appear cynical and immoral. It grates with our rosy view of human reproduction as being, essentially, a loving and cooperative business. It is important to stress that when evolutionists talk of strategies and conflicts of interest, they are talking about cryptic mechanisms that men and women are completely unaware of and over which they can have no conscious control. It is not a woman's will that contests the sacrifice of her store of fats and carbohydrates to support the baby growing inside her—it is the way that evolution has attempted to reconcile the opposing interests of paternal and maternal genes, and those genes when present in the fetus. For Sarah Robertson, this begins with immune regulation of conception and embryo implantation and is all about quality control, with males cast as unscrupulous salesmen perfectly happy to sell shoddy goods, in the form of their contribution to low-quality or incompatible embryos, and women cast as the scrupulous shopkeeper unwilling to stock her shelves with them.

Robertson says that there is variation between males in the active ingredients in their semen that elicit the receptive immune response in the uterine wall, which increases the chance of any individual male siring a pregnancy. But that very receptivity provides the female with the chance to assess the male's reproductive quality and expel the embryo if it is found wanting. Why shouldn't this element of female choice extend throughout pregnancy? In the same way that a poor-quality embryo, if it successfully implants, will consume vital female resources, at least until it fails and aborts, a developing fetus, consuming ever-more resources, can become a maladaptive drain on any woman should her world turn hostile during her pregnancy. That hostility might include infection, malnutrition due to food scarcity, stress caused by some environmental upheaval—a natural disaster or even war—or her male partner, and often her main provider, deserting her. In that sense, although Robertson does not use the term, the embryo and fetus are in a constant state of probation until birth (and indeed afterward).

Female choice, argues Robertson, extends from the embryo to the fetus and placenta. The maternal immune system is exquisitely poised to switch from benign to hostile in response to environmental fluctuations. "One can see," she says, "a powerful mechanism to expel gestational tissue if the necessity arises." While semen chemistry is important in establishing a benign abundance of regulatory T cells at the outset of pregnancy, she explains, that duty falls to signals from the conceptus and the environment later on. Dendritic cells and the T cells they talk to are acutely sensitive to environmental stressors—signals that the immune system receives from the outside world via the hypothalamic-pituitary-adrenal axis. This linking mechanism between the environment and the mother's immune system, she points out, explains how the immune system can react to acute stress and cause miscarriage. It is consistent with recent studies showing an association between miscarriage and high psychosocial stress perception, reduced progesterone, and a shift to Type I (pro-inflammatory) immunity in women. Acknowledging a quality-control function for the immune system, she contends, might change the commonly held view that in women all pregnancy losses are "pathological." "Instead, perhaps immune-mediated loss in certain circumstances is a normal and valuable aspect of optimal reproductive function." I am sure David Haig would nod in agreement.

So far, there is precious little evidence to support such a provocative and wide-ranging idea, although it is clear that an embryo is particularly vulnerable at key points in gestation like implantation and placental development. And it *is* known that a mother's mental state can communicate itself to the fetus via the HPA axis. Daniel Kruger, from the University of Michigan School of Public Health, trawled year 2000 birth records for over 450 US counties and compared low birth weights and premature births with US Census data that revealed the number of single mothers (therefore the degree of male scarcity) and socioeconomic status. He discovered evidence that when males went missing, it was bad news for the baby and that the stress caused by lack of male support resulted in shorter gestation times and lower birth weights. When males are scarce, he says, it is possible that it leads to subconscious mechanisms that limit maternal investment. Whether these mechanisms can be powerful enough to actually put an end to fetal life remains to be seen.

Meanwhile, there is tangible evidence on the horizon that the brutal logic of evolution-inspired reproductive medicine, which so punctured our rosy picture of pregnancy, may lead to important breakthrough treatments for diseases of pregnancy. Ananth Karumanchi, for instance, has shown that the rise of sFlt1 in a mother's blood markedly precedes the onset of preeclamptic symptoms, which suggests that monitoring sFlt1 levels throughout pregnancy can act as an early warning system. His colleagues have demonstrated, in a small number of preeclamptic women, that removal of sFlt1 by blood dialysis stabilized blood pressure and proteinuria and prolonged pregnancies. Thanks to research collaboration between Sarah Robertson and Gus Dekker in Australia, there are hopes that treatment with TGF-β can help apparent infertility, and Robertson is involved with a biotech company investigating the use of the cytokine GM-CSF in improving blastocyst quality and the prospects for successful implantation. Jan Brosens believes that IL-33 might prove a useful therapy in the treatment of recurrent pregnancy loss.

Evolutionary medicine may take the romance out of procreation, but it is replacing it with a much greater understanding of what have been, until recently, a number of inexplicable, distressing, black-box diseases of pregnancy, which everyone agrees have been badly served by scientific research for too long. The research being done today by Brosens, Robertson, and many others has its roots in evolutionary theory dating back to 1974, when Robert Trivers, arguably the greatest living evolutionary theorist, first introduced the world to parent-offspring conflict theory. It is David Haig who has extended this theory into the realm of diseases of pregnancy. The success story for the application of evolution to medicine is that several of the world's top reproductive biologists are now using it explicitly as the theoretical framework for their research.

THE DOWNSIDE OF UPRIGHT

THE RELATIONSHIP BETWEEN BIPEDALISM AND ORTHOPEDIC ILLNESSES

There's an old story that has been doing the rounds for years concerning the legendary Massachusetts congressman and former Speaker of the House of Representatives, the late Tip O'Neill. A gregarious politician, O'Neill was never afraid of pressing the flesh. But his memory for faces fell woefully short of the numbers of potential voters he frequently greeted on the stump. His canny gambit was to gently put his arm around their shoulders, shake their hand, and quietly ask them, "How's your back?" Since about 80 percent of his audience would have suffered from one form of back pain or another, this "personal concern" was nearly always guaranteed to elicit gasps of awe and admiration—"That's amazing—he remembered me!"

Without doubt, one of the most important features of modern humans is our ability to walk upright. It freed our hands, made us into more efficient hunters and foragers, and allowed our brains to grow. Those feet, hands, and brains rapidly spread across the planet and have changed it, not always for the better, in many profound ways. Throughout the animal kingdom, only birds and humans are true, obligate bipeds, though some other primate species, like gibbons, are bipedal whenever they move on the ground. Most birds hop, but a few, like the ostrich and its relatives, are efficient walkers and runners. But only humans combine true bipedal locomotion with a fully erect spine through which weight bears directly downward, like a plumb line, from head to pelvis, and on to the feet. Some evolutionary anthropologists think there is a very good reason why our form of bipedalism is unique—a one-off experiment—throughout the animal kingdom. They argue that it is because it comes with quite literally crippling costs that manifest themselves today in a horrendous and debilitating spectrum of disease and morbidity of our muscles, tendons, and skeleton, which have all come about because we have turned our spines through ninety degrees into the vertical.

Take Rosalind Michel, for instance. In 1951, in her eleventh year, Rosalind first began to realize that all was not well with her spine. She started getting pains in her lower back and legs, and, disconcertingly for an active child, she began to lose the power in her legs that normally propelled her quite happily up the steep hill to her house from the bus stop on her way home from school. Her doctor passed it off scornfully as "growing pains" and shooed her away to get on with the business of growing up, but her difficulty getting home gradually increased. She was in great pain and had to lie down as soon as she got inside the house, while the pain subsided.

Her mother could stand it no more and took her to see a specialist. "He took one look at my back and said, 'Bend down and touch your toes.' My mother gasped because I just couldn't get my hands below my knees!" The specialist told her he thought that something had slipped in her spine and that she would need an operation. In fact, her lowest lumbar vertebra, L5, instead of being perched on top of her sacrum, had fallen in front of it, dragging the rest of her spine behind it. The condition is known as spondyloptosis. Nobody knows for sure why such a disastrous event can happen, though there may be a genetic component to it. Rosalind put it down to over-exuberant play at school: "I was a child who had a very bad habit of jumping off high walls. I went to a school that had lots of lovely terraces in the gardens, and it was a sport of ours to see who could jump off the highest. So it may have been my own fault!"

In the hospital she first had to endure traction for nine days. They put some pins on her legs and strung them up to try to get the weight of her body to pull her spine into place. It didn't work and so they operated. There were no titanium rods or bolts in those days, so, using bone grafts collected from the wings of her pelvis, they fused the base of her L5 vertebra directly to the front of her sacrum in the hope that they could prevent it slipping any further. She had to spend the next twelve weeks in the hospital, flat on her back in bed, encased in plaster, while the bone graft knitted. She then got up, continued her education, went to medical school, and embarked on a profession as an anesthesiologist, all the while with a surgical fix that, over time, led to a concave bending of the spine, shaped like a reversed capital C—known as a lumbar kyphosis—at just the point where normally the spine should be C-shaped in what is termed a natural lordosis.

Rosalind Michel had joined the ranks of millions of people whose lives are racked by back pain, slipped disks, extreme curvatures of the spine, and discomfort and degeneration that descends from neck to toe. Many physical anthropologists and evolutionary biologists attribute this catalog of damage to the fact that we are the only vertical bipedal species. This tale of woe represents, to them, the huge costs imposed by the redesign of the spine and its appendages for bipedality: the price we pay for walking tall. The pioneer of this idea was Wilton M. Krogman, a forensic anthropologist from the University of Pennsylvania whose encyclopedic knowledge of the human skeleton earned him the nickname the "Bone Detective" from Philadelphia's police force. In exactly the same year, 1951, that Rosalind was lying in a hospital encased in plaster, he wrote a seminal paper, which would have resonated painfully with her, called "The Scars of Human Evolution," which claimed: "We humans are such a hodgepodge and makeshift that the real wonder resides in the fact that we get along as well as we do."

For Krogman, four-legged animals are "the bridge that walks" because the spine operates like a cantilever arch supported on four piers, the legs, with the chest and abdomen suspended underneath. When we rotated the spine through ninety degrees, said Krogman, we lost all those design advantages and the spine had to accommodate itself to vertical weight-bearing by breaking itself up into three arches. We are born with a simple ancestral arch, but as soon as we lift our heads, we begin to develop a C-shaped curve in the neck, called the cervical lordosis, and when we start to walk we develop a C-shaped curve in the lower trunk, called the lumbar lordosis, separated by a curve going in the opposite direction, the thoracic kyphosis, all making the characteristic S-shape of the human spine. To permit all the twisting and bending that human bipeds do, the individual vertebrae became wedge-shaped, with the thinner edge to the rear, rather, explained Krogman, like the segments of a toy snake. This hinge-like mobility comes at a price, particularly when the lower back is under load and one vertebra can slip against the slope of its neighbor.

We reap most of the evil consequences of standing on our hind legs in the area where lower spine, pelvis, and sacrum come together, Krogman pointed out. Our pelvis sits at an angle to the backbone, not in line with it. The iliac bones became shorter and broader and help to support our sagging viscera. The weight-bearing stresses are focused where the iliac bones articulate with the wedge-like sacrum. In humans this sacroiliac articulation has been elongated, which has pushed the sacrum down to a point where it can encroach on the birth canal, narrowing it and making childbirth more difficult. This area, said Krogman, is inherently unstable and the source of much of our lower back pain.

The huge weight of our abdominal viscera, obeying gravity, leaves us susceptible to hernia. The positioning of the heart high above the ground means that venous blood must overcome four feet of gravitational pull to return to it, leaving us open to varicose veins, while the veins at the lower end of our large intestine can become easily congested, leading to hemorrhoids.

Krogman then rushed south to take in the feet. In a blink of evolutionary time, they have changed from being grasping and prehensile into a load and stress-distributing mechanism. "We have brought the big toe in line with rest of the foot, and the combination of the tarsal bones, which make up the heel, ankle joint and instep, now account for half the total length of the foot, compared to one-fifth for the chimpanzee. We have created a rigid arch and two crosswise axes, one through the tarsals, the other through the main bones of the toes." It is, claimed Krogman, a bit of a botched job. He concluded: "Our fallen arches, our bunions, our calluses and our foot miseries generally hark back to the fact that our feet are not yet healed by adaptation and evolutionary selection into really efficient units."

In February 2013, a group of modern Krogmans was singing from the same hymn sheet at a special session of the American Association for the Advancement of Science, entitled, in deference, "The Scars of Human Evolution." Ann Gibbons, the veteran correspondent for *Science* magazine, recalls that Bruce Latimer, from Case Western Reserve University, and a survivor of back surgery, "limped to the podium dangling a twisted human backbone as evidence of real pain" to argue that our vertical stance gives rise to a host of injuries and conditions unique to humans. Our S-shaped spine accumulates stress over time, leading to an exaggeration of its natural curves into swayed backs, hunched backs, and scoliosis—lateral curvature. A particular problem, he said, was the large, C-shaped curvature of our lower back that is necessary to prevent obstruction of the birth canal and balance the upper body over the legs and feet. It is a structure conspicuously vulnerable to wear and tear: "If you take care of it, your spine will get you through to about forty or fifty," she reports him saying. "After that you're on your own!" The twisting motion created by planting one foot in front of the other as we walk, with the arms swinging in the opposite direction, wears out and herniates our inter-vertebral disks over time. Only our species, argues Latimer, regularly suffers from fractured hips, bunions, hernias, fallen arches, torn menisci, shin splints, herniated disks, fractured vertebrae, spondylolysis (damage to a delicate part of the dorsal side of the vertebra caused by over-extending or arching of the back), scoliosis, and kyphosis.

When we walk and run, says Latimer, the forces generated across our joints can approach several multiples of our body weight. This is not too worrying when walking because the fifteen-degree flexion of the knee can absorb a lot of that energy. If we were to bend our knees even more when we walk, we could dissipate even more energy, but, as you will see if you try this yourself, it soon gets very tiring! Unfortunately, when we run, our lower limbs become fully extended at precisely the time we are generating maximum forces. In a chimpanzee, the thighbone, or femur, hangs down straight from the hip, and the bones of the lower leg continue in a straight line below the knee, which means that stress on the knee joint is symmetrical. In humans, because the feet have to be brought directly under the body's center of gravity, the knee joint is angled inward, leading to uneven wear and tear on the cartilage.

Jeremy DeSilva, from Boston University, echoes Krogman's point that evolution has not had enough time to create a perfect human foot. If our feet were well designed, he points out, would podiatry be a billion-dollar industry? An ostrich foot is a much better tool for running. Ostriches have fused their ankle and foot bones into one single, rigid bone, much more like the blade runners of modern Paralympian sport, and have long, thick tendons to store elastic energy during bipedal motion. Ostriches, and their ancestors, have had 250 million years to perfect their solution. We have had a mere 5 million. We humans still have twenty-six bones in each foot because we have descended from tree-living ape ancestors who used all these bones in a highly articulate, muscular foot ideal for grasping branches. Of course, says DeSilva, evolution has strenuously bent the ape foot into its new job of walking and running: "The human foot does its job quite well, and natural selection has fiercely molded the foot into a functional structure that absorbs ground reaction forces, stiffens during the propulsive phase of gait, and even has elastic structures like the arch, and the Achilles tendon, to put a kick in our step."

Nevertheless, he reminds us, we would be hard-pressed to call the design of the human foot "intelligent," because it is not designed at all—it is the product of a myriad small, blind, jerry-rigged modifications to preexisting structures. These modifications, for DeSilva, are the biological equivalent of paper clips and duct tape. Evolution, unlike the engineers who designed the prosthetic running feet we associate with the troubled Paralympian Oscar Pistorius, cannot go back to the drawing board and start with a blank sheet; it has to make do with what is in front of it, and in the case of the human foot that was a highly flexible, flat-footed structure with a long, opposable big toe ideal for wrapping around branches. "Naturally," DeSilva explains, "there are consequences for this jerry-rigged system: ankle sprains, plantar fasciitis, collapsed arches . . ." to which you can throw in the development of heel spurs, inflammation of the plantar aponeurosis (the long ligament connecting the heel to the base of the toes), and the malevolent bunion that can form when the big toe is forced out of alignment.

Back pain is a worldwide epidemic phenomenon. In the United States alone, 20 million people seek treatment from a doctor for back pain every year, and annual spine care costs exceed $85 billion. In the UK, over 81 million working days are lost to back pain every year, and treatment costs to the National Health Service alone top 350 million pounds. This excessive morbidity has given rise to a huge industry geared to the diagnosis and alleviation of the symptoms of back, leg, and foot pain. At the Back Pain Show at London's Olympia exhibition center in July 2014, that industry was on full show: everything from walk-in MRI body scanners to diagnose back conditions, to massagers, chairs, beds and cushions, kinesiology, yoga, alpha-brain-wave-inducing music, micro-current therapy, orthopedic traction, magnet therapy, alternative nutrition for the alleviation of bad backs, and bamboo extract capsules to repair damaged vertebral disks.

Pete May is a typical lower back pain sufferer. He is a sports writer who cannot practice what he preaches without having to stagger to his osteopath straight afterward. Now fifty-three, he has been shouldering back pain since his early twenties, when the twisting and turning of five-a-side football first revealed problems. "I remember once trying to make it to my osteopath, who lives up a long hill in north London, and I had to take a taxi up there, and the osteopath had to give me a lift back down again! I was hobbling down the street! The last time my back went out, I was walking on the coastal path in Cornwall. It was very muddy, I slid, my legs went from under me, and I landed on my bum. The next day I couldn't walk upright."

His osteopath, like a garage mechanic nursing a vintage motor, keeps him going. Typically, Pete's diagnosis is necessarily vague, there's nothing obviously broken and out of place, but it hurts. All that can be said is that he has mechanical low back pain caused by irritation of the delicate facet joints at the rear of the spine. And he will probably never get better.

Certainly, on the face of it, we humans appear to have paid a high price for our deportment. And although it is true, as Bruce Latimer points out, that the longer we live, the more likely it is that we will succumb to problems associated with our back and lower extremities, there is plenty of evidence for a number of serious medical problems associated with the musculo-skeleton that are not strictly related to aging. If being bipedal has always carried a high price, it seems only reasonable to argue that there must have been strong compensating advantages to two-legged locomotion that counter the downside of upright. Not surprisingly, there has been no shortage of theories, most of them assuming that the theater for the evolution of bipedalism was hot, open, African savanna grassland to which our ancestor species had transitioned as the forests shrank. However, none of these theories have proved ultimately convincing.

The first to be knocked on the head was the "watching out" hypothesis that suggested bipedalism would have allowed individuals to stand up and look out over the long grasses. However, field research on chimpanzees and gelada baboons suggests they rarely, if ever, rise up to scan the horizon. Don Johanson, the famous paleoanthropologist who discovered Lucy, also witheringly points out that for an ancestral hominin species like *Australopithecus afarensis*, which was only three and a half feet tall, weighed only sixty pounds, and was not fleet of foot, to thus advertise its presence to a range of fast and powerful predators would have been a distinctly bad idea!

The "freeing up the hands" hypothesis proved irresistible for many years with various authors suggesting tool using, weapon handling, food gathering, and self-defense as likely prime movers. But Johanson notes that there is over a million-year gap between those former two activities and the emergence of bipedalism, while chimpanzees and other apes gather food and fight without resorting to obligate bipedalism. Others have suggested that bipedalism freed up female hands to carry infants who had lost the ability to grip their mothers' pelts because of a remorseless trend toward hairlessness. However, two scientists from Manchester University tested the hypothesis by measuring the energy use in fit, young women who were asked to carry symmetrical ten-kilogram weights either in a vest or two dumbbells, versus an asymmetric displacement of the same load as would be the case when carrying a baby on one hip or in the arms. Carrying the asymmetric load consumed so much extra energy they concluded it was highly unlikely that the need to carry naked babies selected for bipedalism.

In similar brusque fashion, Johanson disposes of the "thermoregulation" hypothesis that suggests we became upright to reduce the heat load from direct solar radiation. Why not rest up in the shade of a tree at the height of the day, he asks. Finally, the "aquatic ape" hypothesis proposes that human hairlessness, efficient sweating, and subcutaneous fat more closely fitted an aquatic phase in our evolution, rather than adaptation for life on the savannas. This theory has been efficiently disposed of elsewhere, but a clever "watered-down" version by the German anthropologist Carsten Niemitz hypothesizes that, by wading in thigh-deep water, early biped human ancestors, still tottering a bit, could have benefited from the additional support of a body of water while they found their feet. However, they would not have benefited from the very high likely costs of predation in waterways teeming with crocodiles, hippos, snakes, and giant otters!

We have, says Johanson, been asking the wrong question. There is no monolithic answer to "Why did our ancestors become upright?" Instead we should be asking: "What, when, and where were the advantages for early hominids that resulted in a behavioral change from quadrupedalism to bipedalism?" The answers to these questions have proved somewhat intractable due to the scarcity of fossilized remains of our ancestors and the difficulties, until the use of relatively modern isotopic techniques, in re-creating the environments in which they would have lived. But a picture is gradually emerging, and it contains a few surprises.

All the early species in the genus *Homo*—like *erectus*, *habilis*, and *heidelbergensis*—were full-fledged bipeds. But how much earlier did our ancestors convert to bipedalism? The well-known "pavement" of *Australopithecus afarensis* footprints, discovered at Laetoli in present-day Tanzania, proves beyond reasonable doubt that some australopithecines were sophisticated bipeds at least 3.5 million years ago. The footprints were very human-like; they had an arch, the big toe had been brought into line with the rest of the foot, and there was evidence for a heel bone much bigger than that of a chimpanzee. Indeed, Bruce Latimer has joked that they would not raise eyebrows if encountered on a Floridian beach today! Robin Crompton and his colleagues from Liverpool University used fMRI imaging to make a 3-D impression of the Laetoli footprints that they compared to those from modern humans. Their computer simulation gave estimates of footprint pressures across the foot and allowed them to predict what type of gait would match with the Laetoli prints. They concluded that, far from walking in a crouched posture like an ape, *A. afarensis* walked like a human, fully upright and with the stride driven by the front of the foot, particularly the big toe.

Could they climb trees? Jeremy DeSilva, working from a number of fragments of australopithecine feet, has concluded that most members of the genus had a very square human like ankle joint, unlike chimps, whose ankle joint is trapezoid and allows for a huge amount of flexion at the ankle. When chimps shinny up trees, they bend their feet right up until they are practically touching the shinbone. Australopithecines couldn't do this and would have been more ungainly climbers. So, while it is possible that australopithecines labored up trees for fruit gathering, escape, or to make night nests, their main home seems to have been on the ground. That having been said, there are a few australopithecine anomalies. In 2012 Yohannes Haile-Selassie, Bruce Latimer, and colleagues presented evidence of fragments of an australopithecine foot, dating to 3.4 million years ago. It has yet to be identified by species name, but it is possible that it belongs to an earlier species, *Australopithecus ramidus*, that had persisted alongside *Austrolopithecus afarensis*. There was enough of the new specimen to show that it had an opposable big toe. The researchers conclude that this species, code-named BRT-VP-2/73, was more nimble in trees than on the ground. This suggests that the march to bipedalism was not a straight line but had terrestrial bipeds coexisting with more arboreal members of the same genus.

Scientists were quick to point out that the new find shares many features with an earlier hominin species, *Ardipithecus ramidus*, that existed at least 4.4 million years ago, perilously close to the conventional date for the divergence of human and chimp ancestors from a common ancestor. "Ardi" is a game changer that has forced a major revision of our understanding of the origins of bipedalism. We are all familiar with the infamous frieze of human evolution from a knuckle-walking chimpanzee to an ungainly, slouching australopithecine, and finally, via *Homo erectus*, to the graceful, upright *Homo sapiens*. According to a growing number of paleoanthropologists, that cartoon can now be safely thrown into the trash. Ardi was clearly a biped; her pelvis much more closely resembles the broad, flat paddles of the human and australopithecine pelvises. Together with fragments of a femur, a few crushed ribs, and the posterior part of a thoracic vertebra, they make clear that Ardi stood upright. But she had long, slender feet with a large opposable big toe.

C. Owen Lovejoy, Tim White, and their colleagues have concluded that Ardi was a biped adapted for moving safely in trees, with feet capable of grasping small branches for support as she "walked" through the tree canopy. Yet she would also have been capable of walking upright on the ground, though not with the efficiency of the later australopithecines. "*Ardipithecus ramidus* thus now provides evidence on the long-sought locomotor transition from arboreal life to habitual terrestrial bipedality," they asserted. And then, rather more tartly, "As a consequence, explications of the emergence of bipedality based on observations made of African ape locomotion no longer constitute a useful paradigm."

Some anthropologists have long maintained that because humans, and their direct ancestors, share the same shaped wrist and knuckle bones with chimps and gorillas, they must have descended from a knuckle-walker. However, Tracy Kivell and Dan Schmitt have shown not only that chimps and gorillas use different parts of their hand and wrist for "knuckle-walking," but, by extending their observations across a wide spectrum of apes and monkeys, that many species with wrist and knuckle bones similar to chimps do not, in fact, knuckle-walk, and, conversely, that many species who do not share the same bone morphology as chimps do.

The discovery of Ardi has reignited interest in a number of oddball hominins from further back in time. It is beginning to look as if these fringe characters in the human evolution story may have to be moved center stage because they may represent a number of different experiments with bipedalism in the forest canopy among our direct ancestors that go back some 9 million years. Take *Oreopithecus*, for example. This animal, whose fossils bear a strong resemblance to *Ardipithecus* but is not related to it, was a native of Sicily and Sardinia, in the middle of the Mediterranean. It had a broad chest, long, slender fingers and toes, highly flexible joints, and a lumbar lordosis—which meant it could stand fully upright. It would have been a skillful mover in trees and would have been capable of some degree of ungainly walking. Its big toe was turned out at a 100-degree angle to the rest of the foot so that the foot acted rather like a tripod. It would have had no natural enemies in its island refuge until land bridges formed to the African mainland about 6.5 million years ago. Fierce predators moved in, and *Oreopithecus* was soon wiped out.

In 2001 the French paleoanthropologist Michel Brunet unearthed a battered and distorted braincase in Chad that he named *Sahelanthropus* (though its nickname is Toumai, "hope of life"). The skull has a mixture of human-like and ape-like features, but Brunet is convinced its owner was bipedal because of the forward position of the foramen magna, the point at which the spinal cord exits the skull. In 2000 an early hominoid was discovered in the Tugen hills of Kenya and named *Orrorin*. It had small teeth, relative to both australopithecines and *Ardipithecus*, and the point at which the femur articulates with the hip, the trochanter, is quite similar to humans. This strongly suggests that although it would have been an adept tree dweller, it also was bipedal.

Even though it has not yet been decided whether any of these early bipeds are our direct ancestors, it is clear that experiments with bipedalism were far from rare among hominoids (the group that contains the great apes and ourselves). According to Robin Crompton, this means that it is extremely unlikely that human bipedalism evolved from chimp-like knuckle-walking. Our ancestors would have had to evolve from earlier primates that were quadrupedal, into so-called crown hominoids (like *Sahelanthropus*, *Orrorin*, etc.), who were bipedal, back into species like *Pan* and *Gorilla*, which were quadrupeds, and, finally, with the australopithecines and the genus *Homo*, back again into fully erect bipeds! Crompton and others have reached for Occam's razor. They far prefer a more parsimonious scenario that has us descending from committed bipeds earlier in the hominoid family tree, to a bipedal common ancestor with the ancestors of chimpanzees. Our hominin line then continued with the bipedal project while chimps and gorillas went the other way—evolving from habitual bipedalism into their own peculiar forms of quadrupedalism, with occasional use of bipedalism, or even tripedalism for foraging, fighting, display, and scrambling into the forest canopy.

Crompton's theory seems, at first sight, paradoxical, but this is how he defends it:

> The logical conclusion from the fossil, environmental, and experimental evidence is that upright, straight-legged walking originally evolved as an adaptation to tree-dwelling. But how can this be? Apes and monkeys nearly all increase the flexion of their fore and hind limb joints when moving on small, flexible branches. This acts to lower the center of gravity and hence reduces the chance of toppling, and also reduces branch vibration. Our hypothesis is that bipedal

> locomotion in the trees might actually be advantageous for arboreal apes on the thin terminal branches because their long prehensile toes can grip multiple small branches and thus maximize stability, while freeing one or both hands for balance, feeding or weight transfer while crossing gaps in what primatologists call the "terminal branch niche." We found that when orangutans are moving on the single, largest, most stable supports, they used quadrupedal gaits; when they move on slightly smaller supports they use suspension; but when they move on multiple really small supports they use bipedalism and straighten their legs in a human-like way.

Erect spines have been around in our lineage far longer than we thought. By about 4 million years ago, the majority of australopithecines were reasonably efficient, committed bipeds. Their lumbar vertebrae, in number and wedge shape, were similar to us. They stood tall and did not slouch or lumber about. Their chests, like ours, were more barrel-shaped than chimp chests, where the rib cage is more conical to allow for greater room at the bottom to house their large guts, and less room at the top to allow for the powerful musculature that permits them to fully suspend themselves in trees. Chimps also have fewer lumbar vertebrae than we do, so they could never stand truly upright.

This revolutionary idea about the origins of human bipedalism has very recently been given support because of a drastic revision of the likely date of the split of human and chimp ancestors from a common ancestor. Until 2013 this was thought to have been between 4 and 6 million years ago, which would have meant, because of the discovery of Ardi, that hominins (the branch of apes that gave rise exclusively to us) must have been evolving bipedalism at lightning speed. Now scientists have recalibrated the molecular clock by which they measure ape divergence. By arriving at new estimates for the length of each generation of our deep ancestors in years and for the number of new mutations that arise with each generation, they have pushed back the date for the divergence of chimpanzees and humans, even by the most conservative estimate, to about 7.5 million years. This has immediately swept ancient bipeds like *Sahelanthropus* and *Orrorin* into the human family tree, making chimps and gorillas the outliers who have been going their separate ways for much longer than previously supposed.

Although australopithecine locomotion was not as energetically economic as ours, they would have used less energy than chimps do when walking on two feet. Michael Sockol, David Raichlen, and Herman Pontzer have done the numbers. They found that because of their short hind limbs, bipedal chimps are forced to generate much greater force every time their feet hit the ground. This translates into energy costs. Their awkward bent-hip/bent-knee gait places the body's center of mass forward of the hip joint, which requires a great deal of work in the hip extensor muscles to counteract it. The crouched posture also means they use far more muscle power around the knee. It is far more costly for chimpanzees to walk and run, either on four or two legs, than it is for humans—overall locomotion costs for modern humans are 75 percent less than for chimps. These results suggest that any early hominin that reaped some energy savings through minor increases in hip extension or leg length would have given natural selection something to bite on. That could have translated into more efficient foraging in environments where foodstuffs were occasionally scarce and sparse.

Don Johanson believes this set the stage for the gradual movement of our ancestors farther afield:

> I believe that our ancestors, pre-Lucy, became upright in the protection of the forests, a familiar environment in which they faced fewer dangers. Then, armed with bipedalism, they were pre-adapted to move onto the savannas and expand their territory. Once we moved into more open environments, we brought a whole package of advantages with us. Our hands were free to make and use tools, we could walk long distances to collect and carry food, we could look over tall grass if we needed to, and so on. Bipedalism was the behavioral innovation that led the way to making everything possible for our evolution, even if it is still not perfected. Humans continue to suffer from fallen arches, hernias, lower back pain, and other bad side effects of bipedalism. But it was still the single step in our ancient past that led to the tool-making, brain enlargement, and intelligence that have led to our preeminence on the planet today.

So, is the human spine fit for purpose or, as a number of evolutionists would have it, a flawed, jerry-rigged solution to freeing up the hands that has plagued us ever since? Canvassing a number of orthopedic surgeons and comparative anatomists has persuaded me that we need to take a more nuanced approach and that the human spine has been rather too maligned by evolutionary anthropologists. Take lower back pain, for instance. John O'Dowd is a consultant orthopedic surgeon at St. Thomas' Hospital, London, and a world expert on backs. He doesn't dispute that lower back pain is ubiquitous at high levels worldwide but explains that, for a doctor, it is notoriously difficult to pin down. The problem is that you can have damage to the spine without lower back pain, and you can have lower back pain but no obvious signs of pathology. Many studies suggest that back pain is as common in people with normal spines as it is in people with degenerate spines. It may be that the source of the pain is in some muscle or ligament that is difficult to image on a scanner or, O'Dowd notes, the pain may be in the brain rather than the back. Although real psychosomatic back pain is very unusual, psychiatry does come into it. Imagine you live with a partner who has to look after you year in and year out because you are debilitated by back pain. A doctor then comes along and says there isn't any degeneration in the back. That partner relationship is going to be threatened. Equally, people "know" what is going to be on their scan before they have it because they know what they have to see to validate their back pain. The worst thing for these people is to see that their back appears normal!

You might think that Rosalind Michel, whose spine dislodged completely from her sacrum at age twelve, would have endured a lifetime of misery after surgery, primitive by modern standards, had fused her lowest lumbar vertebra to the front of the sacrum. But her surgical "fix" lasted her for fifty years. Her career was completely unaffected, as was her ability to have children. The presence of the human sacrum with its ventrally pointing tip has always posed a problem for human mothers because it constricts the birth canal. Rosalind's fused vertebra had made the canal even more compromised. Nevertheless: "I had two children along the way which I was told I would have to have Cesarean section for, but they both managed to wiggle round the bend!"

Finally, by her late sixties, problems did begin to surface. Because the surgery had created a lumbar kyphosis, instead of a lordosis, she became gradually more and more bent over at the waist, which meant she had to extend her neck to look ahead of her, a bit like a tortoise. The lack of a normal S-shaped curve in her back gave her balance problems and made walking difficult and tiring. Eventually she elected to have an extension osteotomy that took a V-shaped wedge out of the back of her second lumbar vertebra to straighten her up. The aftermath of the operation was unpleasantly painful as all the muscles, ligaments, and tendons of her back, legs, and abdomen adjusted to her new back alignment. Five years after the operation, she feels she is starting to go downhill again, but this is by her own exacting standards. She recently did the coast-to-coast walk across the UK from Cumbria to the North Sea, taking in a hundred miles and several mountain ranges, and she can still travel ten miles a day, for a week at a time, on walking holidays. She rather hopes that the present "fix" will see her out!

O'Dowd thinks the human spine is brilliantly designed for upright load. For most people, it is fit for purpose for four-score years and is immensely strong, flexible, and resilient even when heavily compromised. For him, it works well beyond its design brief. Comparative anatomists tend to agree with him and, remarkably, to argue that there is nothing particularly singular about the human erect spine and that many of the problems we have with it we share with animal species that are quadrupedal.

One evolutionary novelty in the human spine is its unique S-shaped curve—a convex lordosis in the lumbar region and a convex lordosis in the neck, separated by a concave kyphosis at the chest. Most explanations for this curve point out that it is necessary to allow us to stand upright. However, when you think about it, a stiff, vertical rod would do the same job. For Mike Adams, professor of comparative anatomy at the University of Bristol, the secret to the S-shaped curve lies in its value as a shock absorber. The S-bends resist compressive stress that would otherwise inflict terrible loads on an otherwise very stiff structure. Whenever we move, the shape of these curves changes very slightly just like a bedspring. They are constantly under tension and so readily absorb energy. But this spring-like deformation of the spine can only be slight—the large paraspinal muscles, running parallel to the spine, stop those bends from going too far.

Of course that S-shape can become a liability. Swaybacks and hunchbacks are legion. But, without doubt, problems, when they arise, are most likely in our lower back—our lumbar region. Often this is attributed to the compressive loading of the spine, the vertical force coming down the spine perpendicular to the inter-vertebral disks. That force is partly due to gravity, hence greater at the bottom than at the top, but it is also due to muscle tension in the paraspinal muscles that are always under tension to stabilize us in the upright position. They apply just as much compressive force as gravity, and much more during heavy lifting tasks. The human spine is adapted to cope, to some extent, with this loading. The lower vertebrae are bigger and stronger, and, since stress is force per unit area, compressive stress therefore remains equal at any one point as you go down the spine. What really causes the pathology of the lower spine, claims Adams, is bending, not compression.

Adams's former colleague Donal McNally, now professor of anatomy at the University of Nottingham, agrees and points out that muscular compression of the spine is just as great in quadrupeds as it is in us. Just imagine, he says, a cheetah in full cry after a gazelle. The vivid acceleration, top speed, and twisting and turning involved in the chase cause the paraspinal muscles to work overtime, and this imposes massive loads on the spine. Spasm in these paraspinal muscles is immensely strong and potentially destructive. For instance, pigs are bred for bacon and so they have massive back muscles. Sometimes, when the pig is stunned at the slaughterhouse, the stun gun is not used correctly. It can throw the back muscles into a death spasm that is so violent it can crush vertebrae. They are often found in pieces in the meat after butchery. Even humans, in the grip of an epileptic seizure, can crush their vertebrae.

Although it may seem surprising, changing the orientation of the spine from horizontal to vertical makes little difference to the compressive loads that the spine experiences because they come from muscle action not gravity. For example, the lumbar spine of an average man experiences compressive loads of about two thousand newtons when doing a heavy task, and, of this, perhaps only three hundred newtons is due to the weight of the arms, head, and chest bearing downward. The human spine is loaded in much the same way as quadrupeds, says McNally. The original "design" of the spine and the design spec really isn't changed by bipedalism. In fact, notes McNally, if anything, humans are uniquely adapted to reducing these muscle loads because our bodies are very much flattened in an anterior/posterior direction (compared, say, to the barrel chest of a gorilla), and that helps enormously to reduce these bending moments.

The reason why paraspinal muscles exert such big compressive loads is explained by biomechanics. If you want to reduce muscle loading during bending and twisting of the spine, ideally you would have these muscles operating at a distance, on very long levers, but this is not anatomically possible. Paraspinal muscles, as their name implies, run parallel to the spine and therefore operate on very short levers, hence the increased compressive force. Imagine bending over to pick up a fifty-pound weight. This asymmetric loading on the spine has to be counterbalanced by the paraspinal muscles. But the lever of your arm and hand carrying the weight is easily some forty or fifty centimeters from your spine, whereas the lever of your back muscles is less than a tenth of that. So the muscle force in the paraspinal muscles has to be ten times the weight you are carrying. What does your back in is the bending we apply at the same time we are compressing it. When we do things like heavy lifting or gardening, we bend forward and manipulate things on big lever arms, compensating by contracting our back muscles and inevitably applying compressive forces to the vertebrae and inter-vertebral disks. This changes the geometry of the lower lumbar disks. Thankfully, they have evolved a wedge shape that can undergo up to twenty-five degrees of movement, but, nevertheless, under these sorts of strains, problems can ensue.

What evolution *has* provided us with, explains McNally, is a musculo-skeleton that is very plastic. It can change over time—functionally remodel—in order to cope with the strains imposed on it. Muscles, bones, and soft tissues become stronger if persistently loaded. This is one of the reasons why back pain is not a particular problem for people who regularly do jobs requiring heavy lifting. Their backs have remodeled to cope. It is the weekend warriors who suddenly decide to dig over the garden one Sunday afternoon who get back pain associated with loading.

Many of the problems that plague the human spine also show up in quadrupeds, McNally points out, especially if they are particularly athletic animals like horses or greyhounds, or if the animal lives long enough. And with improved veterinary care over the past few decades, we see many more geriatric animals. Chondrodystrophic breeds of dog—like dachshunds, Pekinese, beagles, and poodles—which suffer from a disorder of cartilage formation, can show disk degeneration within the first eighteen months of life, and even non-chondrodystrophic dogs like Alsatians, Labrador retrievers, and Doberman pinschers frequently suffer from very human-like disk degeneration by the time they are eight.

The reason we perceive evolution to be at fault in the anatomy of the human spine, claims Adams, is because of its parsimony. The margins for error are always very tight. We have evolved massive lumbar vertebrae with thick, cushioning inter-vertebral disks that are very resilient and resistant to injury. In fact, they are far stronger than the prosthetic disks that orthopedic surgeons use to replace them. The problem is that when they do get injured, it is very difficult to get them to heal. Blood supply to inter-vertebral disks is virtually non-existent because any blood vessels would simply collapse under compressive load. Also cell density inside the cartilage is very low. Without good blood supply to rush new cells to injury sites, they cannot heal and start to degenerate. Evolution, explains Adams, has left us with tissues that are on the edge. And what often tips them over is culture. In Western societies, we have lost flexibility in our lower spines. We sit in chairs all day and lose mobility. You can make direct comparisons with African or Indian populations, says Adams, where, certainly fifty years ago, they would squat a lot of the time. They had similar mobility to childhood. If you have a very flexible lumbar spine, your disks can bend backward and forward just like a child's and it is more difficult to injure them. If you lose that, any bending will stress the spine — hence the weekend warrior syndrome.

This injury behavior of the disks helps to explain the worldwide epidemic of back pain. Normal disks simply cannot feel pain. Because they are structures that cushion compressive load on the spine, they are not supplied with blood vessels or nerves. But if you do get a slipped or herniated disk, radial fissures can develop, which allow blood vessels and nerves to peek in. Vital molecules that keep the disk hydrated can leach out, and immunological factors that cause local inflammation can flood in. It is inflammation that causes the pain.

Furthermore, we were never designed to stand stationary and upright for long periods, which today we often do, and that concentrates loading quite severely in the posterior part of the inter-vertebral disk and in the neural arch. When you are standing upright, you get a great deal of load transmission through the little delicate apophysial joints to the rear of the vertebrae. We have all watched our elders palpably losing height as they grow older. Much of this is due to disk degeneration that narrows the inter-vertebral distance and thus puts more stress on the apophyses. These little joints resist less than 10 percent of the compressive force on the spine when you are young, but by the time you get to fifty years old, they are shouldering 20 or 30 percent of the load, especially when you arch your back in a lordotic posture. If you have lost disk height, up to 90 percent of the compressive force is through these very delicate joints. This causes gross bone remodeling, exuberant growth of bone spurs, and osteoarthritis.

But if this is the downside of upright, there is also an upside. Evolution has provided us with a skeleton that is very adaptive in the short term, and the effects of this adaptive remodeling can be huge. On the one hand, Andy Murray and Roger Federer will have 35 percent more bone in their racket arms than you or I do, while, on the other hand, if you take to your bed for six months, you will suffer 15 percent bone loss. Remodeling allows our bodies to supply more mass when it is needed and save on valuable resources by withdrawing mass when it is not. Adaptive remodeling of muscle and bone is one of the reasons why super athletes like Usain Bolt set new world records today that will surely be eclipsed by their successors — the limits are far from being reached.

While I believe we ought to take most of Bruce Latimer's dire prognostications on the health of the human spine with at least one pinch of salt, there are three conditions, unique to humans and their ancestors, where our bipedal stance does have to shoulder some blame. They are osteoporosis, pregnancy and childbirth, and scoliosis.

Osteoporosis is a complex metabolic and hormonal disorder, but it does have a biomechanical component. Bones are composed of two types of tissue: cortical bone, which is hard and dense and forms the perimeter of the bone, and trabecular or cancellous bone, which forms the spongy interior, crisscrossed with spiny structures called trabeculae. We humans have vertebrae that get larger as you go down the spine toward the pelvis, and we also have an extremely large heel bone, or calcaneous. But while the lower vertebrae are bigger, they are less dense and have much more trabecular bone and much less cortical bone than do quadrupeds. The necks of our long bones, at the points where they articulate with other structures, are similarly designed. Trabecular bone will bend when force is applied to it, and that makes it an efficient shock absorber.

However, the very features of trabecular bone that make it such an efficient shock absorber leave it vulnerable to osteoporosis. The more bendy and flexible it is, the thinner, longer, and more widely spaced are the trabeculae. It is also alive—dynamic—because the bone cells, called osteoclasts, that resorb bone and thus reduce its density line the trabeculae. So the very large internal surface area of large trabecular bones has an enormous potential to remodel—change its density—in response to signals from the rest of the body. One of those signals is mechanical—the amount of stress loading the bone regularly has to soak up. Although, as Mike Adams points out, there are warring scientific camps over this issue, one idea is that weakening and underused muscles, as we get old, reduce those loads on our bones. They do not experience the high forces they used to, and so they start to offload density to get lighter. Here is evolution's balancing act at work again—the "use it or lose it" principle. But in the case of an elderly person, it only needs a slip on the proverbial banana peel to cause a major and potentially life-threatening fracture.

At the AAAS session in February 2013, in recognition of Walter Krogman, anthropologist Karen Rosenberg reminded us that his title "The Scars of Evolution" also resonates painfully with modern mothers because the very high level of surgical intervention in childbirth, via Cesarean section, literally leaves them with a large scar for life. While the rate of Cesareans may reflect the over-medicalization of childbirth, she noted, it is also a measure of just how difficult and painful conventional childbirth is as a result of the competing selection pressures on the female pelvis throughout hominin history. The pelvis first adapted for upright walking by narrowing, but then had to secondarily adapt to the fact that, from *Homo erectus* onward, babies were being born bigger and with bigger brains, which needed a widening of the birth canal. A baby gorilla, for instance, weighs a mere 2.7 percent of its mother's weight, but for a human baby that rises to over 6 percent.

The diameter at the inlet of the human birth canal is wider from side to side—hip to hip—than it is front to back, and so the baby's head, as it begins to nudge into the canal, has to turn sideways. As the head travels down the birth canal, progress is hampered by the position of the ischial bones of the pelvis, particularly the ischial spines, and the forward-poking end of the sacrum. The baby has to rotate through ninety degrees to progress any farther. This also allows the shoulders to follow the head into the canal. Later, near the outlet, it has to twist again so that the back of the head is now resting against the pubis. It emerges facing away from its mother, which means that to prevent her from accidentally trapping it in its own umbilical cord or damaging its delicate back through pushing and shoving, the job cannot be safely done without the intervention of a midwife or obstetrician today, or a willing female assistant in prehistory. Even with help at hand, until recently childbirth was a very dangerous business and the leading cause of female and infant mortality.

Rosenberg once argued that this represented a classic example of the flawed and imperfect way that evolution goes about its work. In 2007 she told *National Geographic*'s Jennifer Ackerman: "The result of all these different pressures is a jerry-rigged, unsatisfactory structure. It works, but only marginally. It's definitely not the type of system you would invent if you were designing it. But evolution is a tinkerer, not an engineer; it has to work with yesterday's model." Today Rosenberg's attitude, like mine, is a little more nuanced. Like me, she sees evolution by natural selection as more of an inspired tinkerer. Rather than childbirth being the scar of an imperfect evolutionary process, she says, it is better to view it as an inspired trade-off between two virtually incompatible requirements.

Yet even with pregnancy, evolution has tinkered to good effect. As Katherine Whitcome, together with Liza Shapiro and Dan Lieberman, has shown, there is an adaptation of the lumbar vertebrae unique to human females and their female ancestors. During pregnancy, the body changes its shape because the mass of the developing baby, all eight or nine pounds of it, projecting forward in the womb, alters a woman's center of mass and positions it in front of the hips. This makes walking more and more inefficient, leading to the well-known waddle of a heavily pregnant woman. To some extent, this can be counteracted by muscle action, but it is tiring. Evolution has ameliorated the situation. Our lumbar vertebrae are heavily wedged, allowing us the backward arch of our lumbar spines that allows us to stand fully upright. In men this wedging occurs in L4 and L5, the two lowest lumbar vertebrae, but in women it extends upward to L3. This allows pregnant women to extend their backs up to twenty-eight degrees near full term, which repositions the bulk of their baby back over the hips. It also vastly reduces the shearing forces on the vertebrae that would otherwise occur with such an exaggerated curve. We have noted earlier that pronounced lordosis affects the delicate apophysial bony projections on the rear of the spine because it sends more compressive load through them than occurs with normal upright posture. Women have evolved larger apophysial surfaces to cope with this increased loading during pregnancy. None of these spinal adaptations are seen in chimpanzees, but they are seen in *Australopithecus afarensis*, which has exactly the same sexual dimorphism in lumbar wedging seen in humans. Whitcome suggests this means that early hominin mothers suffered just as much as modern mothers from fatigue and back pain, which could have severely limited their foraging ability and their ability to escape predators. This would have set up strong selection pressures for the lumbar lordoses we see in pregnant women today.

Kyphotic and lordotic curves in our spine are perfectly normal adaptations to walking upright and absorbing stresses, unless they go too far, often when we get old. But there is a lateral or sideways curvature of the spine, scoliosis, that is totally abnormal, occurs with surprising frequency, and is thought to be unique to humans and our immediate ancestors (it can only be induced in quadrupeds with rather horrible experimental surgery). Scoliosis seems to have plagued human spines ever since the days of the first truly efficient bipeds. Anthropologists who have examined the Nariokotome Boy (sometimes known as the Turkana Boy) have identified him as a member of the species *Homo erectus*, some 1.5 million years old—and he had a lateral curvature of the spine. As I sat down to write this chapter, the sensational news arrived of the fortunate discovery of the skeleton of King Richard III underneath a parking lot in the city of Leicester. He had always been depicted as a misshapen hunchback, but now the truth about his ungainly figure could be firmly established—he had a pronounced rightward scoliosis.

Rosalind Jana is a budding writer and blogger who won *Vogue* magazine's writing talent contest in 2011, when she was only fifteen years old. Her teenage years have been traumatic and painful by anyone's standards thanks to the development of a severe scoliosis. She first noticed that a "rib-bump" had appeared on the left lower front of her torso. Her mother had noticed it too: "My mum thought the rib bump was bunched-up fabric at first, but it soon became clear something was properly amiss." An osteopath told her that she had a very mild scoliosis but that it wasn't anything to worry about. Both her father and her cousin have slight scolioses. But the curvature got worse, and she began suffering discomfort. "My torso was contracted and, because I am quite tall and have very long legs, there was this absolute discrepancy between the length of my torso and the length of my legs so I felt physically out of proportion. The rib-cage shifted very prominently out to the left so it looked as if I had a lump coming out of my side, and then my right shoulder-blade started sticking out and by the end it was almost a right-angle, almost wing-like. One of the reasons I know that my spine contorted so rapidly was that I had a growth spurt over the summer—I was eating huge amounts of food—and yet I actually got shorter!"

Over the course of six months, her scoliosis galloped from a fifty-six-degree curvature to eighty degrees. She was in increasing pain. School sports were out of the question, and she had to lie down after her art exam because bending over a table for two hours had been agonizing. She developed a wheeze, which made it hard to walk uphill. She could no longer put off the visit to the orthopedic surgeon she had been thrusting to the back of her mind all year. "And he sat down at his desk and he looked at these X-rays and measured me and said, 'She has an eighty-degree curve—Cobb angle—we usually operate on anything above forty-five.' At which point my mum said, 'If she does want to have surgery how soon could you do it?' We were envisioning weeks to a month, but he said, 'How does next Wednesday sound?'" During surgery, her spine was exposed and the muscles pared away. Titanium pedicle screws were put into the vertebral bodies and a titanium rod was connected up to the screws and rotated. The whole assembly of her spine and rods visibly straightened up.

The vast majority of scoliosis is idiopathic, which is medicalese for "We don't know what causes it." It is a neglected disease that has not received the research attention it deserves. Nevertheless, there are clues. Idiopathic scoliosis usually shows up in children just as they are entering their teens and their bodies are growing very fast, so it is a disease of growth itself, not of degeneration. Mild scoliosis is surprisingly common, the ALSPAC child study, conducted by the University of Bristol, has followed fourteen thousand children from birth to teenage and reports mild scoliosis occurs at a rate of 5 percent, though the occurrence of scoliosis at greater than a ten-degree curvature is much less. It is primarily a disease of girls—there is about a 9:1 ratio of scoliosis between girls and boys. What is it about growing spines that could leave them open to developing lateral curvature, and why, when they start curving, is it invariably to the right? It may be linked to lumbar lordosis, the C-shaped curvature of the spine. A lot of scoliotic girls have pronounced lordoses. Some theories invoke muscle disorder, others a defect in the connective tissue associated with the spine, while others highlight what they call disorganized skeletal growth. Though why such factors come into play only on one side of the spine is not clear. Some researchers suggest asymmetric mechanical forces are at play, which induce a lateral wedging of the vertebrae.

All these proposed mechanisms invoke things going wrong at the local level of the spine and its attachments. But could there be some more fundamental cause? A number of studies have identified chromosomal regions that can be linked to idiopathic scoliosis and one gene in particular, CHD7, which is also involved in a number of critical processes inside cells. Abnormal variants are involved in CHARGE syndrome, for instance, which is often fatal for babies. They suffer from heart defects, mental retardation, deafness, genito-urinary malformation, and scoliosis. CHD7 is thought to interact with a number of embryonic cell signaling molecules that guide symmetric development of the embryo.

Working with this idea of symmetry, French researchers Dominique Rousié and Alain Berthoz have identified problems with the inner ear—vestibular abnormalities—in children with scoliosis. They have balance problems and take longer to walk properly or ride a bike. Human running and walking can be seen as a constant state of falling and correction, because we are constantly balancing upon one foot while the other makes a stride. Not surprisingly, Rousié and Berthoz found abnormalities in the corpus callosum, the great tract of nerve fibers that connects the right and left hemispheres of the brain. In the scoliotic children they have tested, this abnormality results in defective left/right commands from the brain. Other researchers have found asymmetries in the size of other areas of the brain in scoliotic individuals.

Rousié and Berthoz used MRI scanners to measure cranial and facial asymmetry in scoliotic youngsters and discovered significant asymmetry in the location of left and right eye orbits, the development of the nasal septum, and the jaw and cheekbone. In turn, the jaw and cheekbone are linked to the basicranium, the underside of the braincase, which houses the cerebellum (which has also been found to be asymmetric in scoliosis) and the bony labyrinths of the inner ear. When the latter are distorted, they explain, this could result in asymmetry in the way that the otoliths (the tiny particles in the fluid-filled inner ear that trigger the sensory cells that register gravity and acceleration) pass information to the postural system—again causing imbalance. Specifically, they propose that these semi-circular canal abnormalities send abnormal outputs to the vestibulospinal tract, a part of the motor nervous system that coordinates movement. It connects the brain stem, via the spinal cord, with the muscles of the limbs and trunk. Damage to it, for instance, causes problems with gait that are often seen in people with scoliosis. These asymmetrical abnormalities, functioning at a time when the spine is very plastic during rapid growth, could easily set a lateral curve in motion.

Caroline Goldberg, formerly of the orthopedic department at Our Lady's Hospital for Sick Children in Dublin, conducted her own measurements of asymmetry in scoliotic children. She used several features of the palm (the technique is called palmar dermatoglyphics) and compared left and right hands. There were significant differences in the individuals with scoliosis. The phenomenon this technique measures is called fluctuating asymmetry. It is a measure of the extent to which the physical form of an organism departs from bilateral symmetry—a fundamental property of stable embryonic development. Some organisms, or individuals, are more sensitive to environmental factors as they develop, and these can knock them off course, which will show up as departures from bilateral symmetry on any number of anatomical parameters. Goldberg points to work done by German scientists in the early 1990s suggesting that girls had greater fluctuating asymmetry than boys or, as Goldberg drily puts it: "Girls are more unstable than boys—always got a laugh!" Goldberg has plotted this developmental instability in both girls and boys together with growth rates and has found that girls get a double whammy because they are growing fastest at just the time, around age twelve or thirteen, that they are developmentally most unstable. The growth spurt in boys starts later. Although other researchers have criticized her ideas, she maintains that this might explain the huge excess of scoliosis in girls compared to boys. Adolescent girls are also skeletally immature and tend to be unusually tall for their age—as was Rosalind Jana.

Mike Adams agrees that the female adolescent growth spurt is essential to understanding scoliosis. We know, he says, that de-mineralized bone—as we see in osteoporotic older women—creeps, or deforms, when it is under load. The only other period when we see under-mineralized bone is in adolescent girls. Their rapid growth spurt outstrips the ability of their bodies to load minerals into their bones. Vertebrae in this condition can creep or deform a fraction of a degree every few hours.

Although more research should be conducted to clear up the matter, it is clear that genes and asymmetry are at the heart of scoliosis. Local to the spine, uneven growth centers in the vertebrae, nervous system abnormalities down the spinal column, weak muscles or tight ligaments, or any combination of these could amplify these fundamental asymmetries into full-blown scoliosis.

Why don't animals get scoliosis? It may be because, here, gravity *is* involved. In erect spines the force of gravity will amplify any asymmetrical structure in vertebrae, disks, or musculature. Or it may be, notes Goldberg, that natural selection is far less forgiving of scoliosis in a quadruped. "Just imagine, in your mind's eye, a cheetah with a forty-degree spinal curvature trying to chase a gazelle!" she jokes. In humans a scoliotic back, even at a thirty-degree curvature, is just as strong as a straight back and is absolutely no barrier to childbirth. It is no barrier to running either—Usain Bolt has scoliosis. Richard III fought valiantly on Bosworth Field with a seventy-degree curvature until a sword split his skull. Whether or not his mind was as twisted as his back is still a matter of academic dispute! As Rosalind Jana, after all her teenage trials and tribulations, rather poetically puts it: "Above all, I know I have backbone. It may not be straight, but it's strong."

A long, close look at the evolution of our skeleton has allowed us to distinguish between the genuine costs of the evolution of bipedalism and examples where evolution has produced inspired compromises while trying to square various incompatible design briefs. But there is one area where evolutionary biology has become a powerful tool to investigate something that plagues many of us in the modern world—running injuries.

Dan Lieberman is professor of human evolutionary biology at Harvard

University, and he loves to run. He was never outstandingly good at it but, ever since student days, has jogged several miles at least three times a week along the banks of the Charles River and the parts of Cambridge, Massachusetts, near the Harvard University campus. And he does it barefoot, joining a recent running-style craze that takes its inspiration directly from evolution. Lieberman had begun to wonder precisely when our ancestors evolved to be competent runners and when the features of the skeleton and musculature that would have given them this ability evolved. Together with colleagues Dennis Bramble, David Raichlen, and others, he identified *Homo erectus*, which first graced the African savannas some 2 million years ago, as the first hominin able to run effectively.

Biomechanics told them that maintaining stability when running is of prime importance. This is why we have such big backsides—trunk stabilization is achieved by contractions of the gluteus maximus muscle, which first greatly expanded in *Homo erectus*. The semi-circular canals are vital for balance and to detect acceleration and head pitch, and there is evidence that they also first increased in size with *Homo erectus*. But, says Lieberman, during running the heel imparts such a rapid and substantial pitching impulse to the head that it would easily overload the vestibular system were it not for the evolution of a mass damping system by which the head is decoupled from the shoulders so that they can function as linked masses. In chimps the head and shoulders are tightly bound together by the massive trapezius muscle, an adaptation for climbing, but in *Homo erectus*, and us, the trapezius is much smaller and interlocks with a novel feature, first established in *Homo erectus*, of a nuchal ligament running from the back of the head to the cervical vertebrae in the neck. He explains: "When you run, and you pump your hands opposite your legs, that arm has a mass that's about the same as your head. And the inertial force that causes your head to pitch forward also causes your arm—the trailing arm—to fall. Our tiny trapezius, together with the nuchal ligament, forms a mechanical strut between the arm, which is falling down, and the head, which is falling forward, and the arm thus pulls your head back just as it wants to fall forward. It's totally cool!"

Lieberman argues that there are over a dozen other skeletal features that could be taken into account that could have improved walking and running in *Homo erectus*. But making the case that *Homo erectus* was the first efficient runner merely set up a secondary question. Precisely what was the ecological context in which running would have been important to survival? We humans have been lulled into a state of false grandeur by the likes of Usain

Bolt and Mo Farah pounding round athletics tracks. In fact, even Usain Bolt could be outrun by a rabbit and would be no match for an enraged chimpanzee! Over a hundred meters or so, chimps can sprint much faster and are more agile. And as for a cheetah . . . But chimps soon run out of puff, and cheetahs generate so much heat when running flat out that they have to stop after barely one kilometer. Was there any context in which the tortoise could beat the hare? Lieberman soon realized that we, and our ancestors, were never built for speed—but for endurance.

While scarcely able to outrun predators, *Homo erectus* and its descendants would have been able to literally run certain prey animals to death. Any reasonably fit human (and by extrapolation *Homo erectus*) can run for miles at five meters per second, says Lieberman, which is in excess of the speed at which a dog must make the transition from a trot to a gallop. This means that a human can outrun a dog at distances of more than a kilometer. And although a horse can reach a maximum speed of about nine meters per second, it can only sustain 5.8 meters per second for long distances. This means that, theoretically, humans could beat horses over marathon distances. It turns out that this is literally true! Every June since 1980, in the mountainous countryside near to the picturesque Welsh market town of Llanwrtyd Wells, men and horses go head to head over a twenty-two-mile cross-country course. Although the horse usually wins, the finish is sometimes incredibly close, with the horse coming home only a few seconds ahead. It took twenty-five years for a man to beat a horse—Huw Lobb managed it in 2005 in a time of two hours and five minutes, beating the fastest horse by two minutes. Two years later two human competitors thrashed the horses by a full eleven minutes!

Any hair- or fur-covered animal that has to pant in order to cool down is at a disadvantage to humans because they cannot pant and gallop at the same time. Panting involves a rapid succession of short, shallow breaths involving only the upper respiratory system, so it cannot satisfy the demands for oxygen uptake. Panting animals can either pant or gallop—they cannot do both. Humans do not couple respiration with stride and have sweat glands and hairless bodies. One milliliter of water lost through sweating accounts for 580 calories of heat. If a human has access to between one and two liters of water an hour, he or she alone can endurance run at the height of the day. We really are the odd man out, explains Lieberman, because, by and large, nature favors speed over endurance due to the dynamics of predator-prey interactions (think of cheetahs chasing gazelles). The makeup of muscle fibers is also important. Animals built for speed have predominantly Type IIb (fast glycolytic) and Type IIa (fast oxidative) fibers relative to Type I (slow oxidative) fibers. The former fast-twitch fibers produce more force but operate anaerobically and fatigue easily. Slow-twitch fibers have higher aerobic capacity but generate less force. Human leg muscles have around 50 percent of each, but—and here's where the idea of adaptive remodeling of muscles comes in again—endurance training increases the proportion of slow-twitch fibers up to 80 percent.

Persistence hunting by endurance running would have allowed our ancestors to run other large mammals to the point of heat exhaustion when even an incredibly rustic spear or club would have been sufficient to finish them off. It would have been relatively safe and efficient. Put in everyday terms, says Lieberman, running fifteen kilometers to kill a large antelope requires fewer calories than the 1,040 kilocalories consumed from a Big Mac and a medium fries at McDonald's! Since a large antelope weighs more than two hundred kilograms and contains several orders of magnitude more calories than McDonald's can pack into one of its meals, it's easy to see that the payoff would have been worthwhile even if the chances of success were only 50 percent.

Lieberman's arguments for an ecological niche where endurance running would have had great survival value for our ancestors have suddenly proved useful in the modern world. An interview with Lieberman in the EDGE scientific e-zine in 2005 makes this clear. In it, he describes giving a public lecture at Harvard on the evolution of running and being disconcerted at the sight of a large man with a big beard, wearing suspenders, sitting in the front row. Instead of shoes he was wearing socks wrapped to his feet with duct tape! "I remember thinking it was some homeless guy from Harvard Square who had just come in out of the storm. But it turned out he was a Harvard graduate who lived in Jamaica Plain and ran a bicycle shop. He came up to me afterward and he said, 'You know, I love running and I hate wearing shoes, and I'm a barefoot runner. In fact, I just don't like shoes. Humans obviously evolved to run barefoot. Am I weird or am I normal?' I thought what a great question."

At the time, Lieberman was suffering from plantar fasciitis, a painful swelling of the connective tissue in the arch of the foot. Could he learn anything from his eccentric acquaintance that would help him run more safely? "We brought Jeffrey into the lab and ran him across the force plate. He ran in this perfectly beautiful, light and gentle way. Most of us land on our heels—

we wear these big, cushioned running shoes with lots of support, and we slam into the ground on our heels. But Jeffrey didn't run like that. He landed on the ball of his foot and then his heel touched down gently. He had no impact peak. It suddenly occurred to me that when this guy, barefoot Jeffrey, ran across our force plate, that he must be normal and that I must be abnormal in the way I'm running in my stupid running shoes."

The lightbulb flashed. Jeffrey was running exactly the way all our ancestors would have run—either barefoot or with very simple moccasins. The medical problem among runners is enormous. Despite over thirty years of advancement in sports footwear—the cushioned heels, ankle support, and firm arches from the likes of Nike and Adidas—between 30 and 70 percent of runners incur repetitive stress injuries every year. The trouble is that there is no agreement as to where to place the blame. Could evolutionary biology turn into a forensic laboratory to help resolve the argument between the conventional "high-tech" camp and the barefoot runners?

The former camp maintains that running is inherently injurious, that the risks of injury are made worse by biomechanical abnormality, sedentary lifestyles, training errors, or simply unfriendly modern running surfaces like hard pavements. The barefoot runners claim that injuries result from poor running style and that modern, cushioned footwear blunts the proprioceptive feedback our brains receive, via the nervous system, from the soles of our feet, which would otherwise force us to correct the way we run. As Lieberman puts it: "The human body was adapted to running in a barefoot style, whose kinematic characteristics generate less forceful impact peaks, which uses more proprioception and which may strengthen the feet. I hypothesize that these factors may help runners avoid injury, regardless of whether they are wearing shoes. Put in simple terms: How one runs probably is more important than what is on one's feet, but what is on one's feet may affect how one runs."

It may be, explains Lieberman, that there is a crucial mismatch between the way our bodies were designed to run in bare feet and the way we run today in modern footwear. Barefoot running gives efficient sensory feedback from the soles of our feet that signals hardness, roughness, and unevenness of the ground and gives our central nervous system the information it needs to make instantaneous decisions that avoid injury. Modern sports footwear cushions accumulative impact forces of which we are sublimely unaware until we wake up one day in searing pain. Furthermore, he notes, "There is reason to hypothesize that shoes can contribute to weak and inflexible feet, especially during childhood when the foot is growing." Shoes with stiff soles, arch supports, and features that control pronation and other movements, he says, may prevent muscles and bones from adapting to stresses that used to be normal. Just as food processing leads to low chewing forces and weak jaw muscles, individuals who grow up wearing highly supportive shoes may develop abnormally weak feet, especially in the muscles of the longitudinal arch. Such weakness may limit the foot's ability to provide stability and other key functions. "This hypothesis never has been tested rigorously, but unshod populations are reported to have less variation in arch form, including a lower percentage of *pes planus* (flat feet), and a lower frequency of other foot abnormalities."

Lieberman considers Lee Saxby to be the world's most formidable barefoot running coach. Everything he teaches is informed by Lieberman's applied evolutionary biology. I attended a daylong teach-in at Saxby's cozy gym in Clerkenwell, central London, to see evolutionary biology literally strut its stuff. In one corner of the gym, a life-size cardboard cutout of Charles Darwin looked sternly and thoughtfully down on the proceedings. Opposite him hung a doleful human skeleton. In another corner a runner was padding rhythmically away on a treadmill while a video camera, mounted at right angles, recorded every movement. Saxby's gym was crowded with a mixture of running coaches, fitness trainers, Pilates teachers, and biomechanics scientists. All of them run a great deal, and most of them have suffered from the usual range of running injuries, including plantar fasciitis, Morton's syndrome, runner's knee, and tendinitis. For instance, Dominic was suffering from recurring Achilles tendonitis and had orthotic supports in his running shoes, while Louise had suffered hip and knee problems and, as a long-distance runner, was keen to avoid surgery on her troublesome knee. All were keen to practice what the acknowledged guru of barefoot running preaches.

They began by looking at a video Lee had made of runners passing a camera in the New York City Marathon. The vast majority of them were well-shod in chunky running shoes, and they were hitting the ground at every stride with a pronounced heel strike. In fact it became obvious that the fatter the heel cushion, the heavier they plonked their heels down. How did the runners gathered in his running clinic fare? The group assessed the videos of every member's running technique—scoring them one to three. A score of one denoted poor posture and a clear heel strike. Hips tended to be bent, meaning 100 percent of the impact ground forces travel up through the heel, knee, and hip. A score of two denoted a flat-footed, flapping foot strike accompanied by poor posture—a flailing leg, a slight forward stoop, or a backside hanging out to dry. The ground impact force this generates is still large and travels up to the hip. This is unskilled barefoot running. A score of three denoted excellent barefoot running technique: short, rapid footfall on the ball of the foot and a relaxed body held vertically so that foot, knee, and hip form a straight vertical line. Saxby reminded everybody that Dan Lieberman, in his long-term study of fifty elite athletes from Harvard's track team, found that injuries were over twice as likely when heel strike was employed. A simple graph of ground reaction forces (measured in human body weights) against running technique clearly shows a sharp peak of over three human body weights when the heel strikes the ground first, a less intense peak when some degree of front-foot strike is used, and a low twin peak, which is the signature of good barefoot style.

Good running, explains Saxby, is all about good movement. Gravity, ground reaction force, elastic recoil, and conservation of energy are all involved. His advice runs counter to the recommendations from most of the conventional fitness industry. Much athlete training places stress on muscle building, even though it is useless to runners. For running, the elastic recoil in tendons is vital. The foot, especially the arch and the Achilles tendon, are great shock absorbers that release stored impact energy like an elastic band at the lift-off point in the running stride. Between them, arch and Achilles absorb 50 percent of the impact energy of footfall, releasing it to aid running movement. Hit the ground with the heel, and all that energy storage and reuse is lost and instead travels up to the knee and hip, where it can cause accumulative damage. If you want to experience the joy of elastic recoil firsthand, try the following: Find a metronome and set it going at a 180 beats per minute. Then jump up and down on the spot at the same frequency, landing mainly on the ball of your foot but also lightly kissing the ground with your heel. You will find it effortless. Now slow the metronome to sixty beats a minute—the frequency of footfall for most joggers. Immediately you are aware of the fact that you are now heavily using your leg muscles—particularly your calf muscles—to maintain jumping. One hundred eighty steps per minute is the frequency used by all elite runners, together with endurance runners from specialized traditional societies.

But good barefoot-style running is far more than just making sure that your foot strikes the ground frequently and at the front. At low speeds or on soft going, your body will tend to tell you to heel strike, and it should transition to a front-foot strike as you increase speed. Heavier people will need to transition earlier than lightly built runners. But this will only happen if the brain is getting the right proprioceptive sensory feedback from the foot. Cushioned heels continually confuse the brain about the actual levels of ground reaction force. White noise in—rubbish out. Feet will take care of themselves if they can sense the ground.

The worst thing a runner can do is to assume the unskilled front-foot technique for which our clinic members scored two. Get stuck here and you will quickly develop tendinitis and fasciitis. This is because your posture is wrong, and posture means getting your center of gravity directly over the supporting features of pelvis and legs. As babies learn to stand, then walk, and finally to run, so adult runners have to repeat ontogeny.

A simple standing test became a sobering experience—and I was first up for it. I had to stand barefoot on a force plate, facing a television screen on which a digital graphics image representing the force generated by different parts of my feet on the ground was displayed. To my horror, the graphics showed that I wasn't even standing upright! One foot was displacing pressure very differently than the other, and both feet were concentrating pressure upon the heel. I encountered great difficulty, even with visual feedback, in correcting my posture so that I was standing properly in such a way that the pressure was evenly displaced over heel and ball of foot and where the pressure high spot was actually underneath the big toe. To my surprise, however, I was not alone. Among all the elite sportspersons surrounding me, none exhibited perfect posture on the force plate—all had to work hard to correct their posture. Everybody had to relearn the basics of posture before they were able to progress to perfect their barefoot running technique.

The experience of using evolutionary insights to throw light on a modern medical syndrome has turned Dan Lieberman and his friends and colleagues into full-fledged evolutionary medics. "What I'm doing is part of a slow, gradual, very barely growing movement in evolutionary biology, which we hope will become part of a larger movement in science in general. And that is using evolution to give us insights into medicine and how we use our bodies. . . . I realized that by studying the evolution of the human body, we could address problems that seem to have been intractable. The barefoot running story, as far as I know, is a perfect example of how taking an evolutionary approach to the body gives us insights about how to better use our bodies."

DIY EYE

HOW DEVELOPMENTAL BIOLOGY CURES BLINDNESS AND REBUTS CREATIONISM

Explanations for the evolution of the eye have long represented one of the more ticklish problems in biology. Because of the complexity of its many interdependent parts, evolutionary skeptics, like the proponents of intelligent design and creationism, have frequently targeted it. They argue that the eye is a fine example of irreducible complexity and could not possibly have evolved by acknowledged Darwinian processes of stepwise mutation and selection. In the opposite direction, a satisfying Darwinian explanation for the evolution of such a complex organ does much to bolster acceptance of evolution in general. In this chapter I want to show how accumulating evidence from a number of inspired biologists puts the question of the evolution of the eye beyond any reasonable doubt while simultaneously showing how recent dramatic advances in developmental biology are helping eye specialists design new medical technologies to mend diseased eyes and restore vision.

The summer of 1992 was the proudest moment in Tim Reddish's life. He won the silver medal for the hundred-meter butterfly and the bronze medal for the hundred-meter freestyle at the Barcelona Paralympics. He was thirty-five years old and totally blind. It was the beginning of a glittering swimming career that brought him a total of forty-three medals in international and Paralympic competition and culminated in his chairmanship of the British Paralympic Association and board membership of the organizing committee for the 2012 Olympic Games in London.

Tim suffers from retinitis pigmentosa, a hereditary degenerative eye condition that involves the accumulative loss of photoreceptor cells in the retina until all peripheral vision is destroyed. It leaves sufferers like Tim with only one or two degrees of tunnel vision. For Tim, the degenerative process began at school age, but it was years before the condition was diagnosed. He just thought he was an awkward boy. He wore spectacles at school and was always clumsily bumping into things. "I just thought, 'I'm always doing things at a hundred miles an hour and not looking where I'm going, and I've clattered!'" In the UK in the 1980s, opticians did not routinely test for retinitis, and so Tim's deteriorating condition was not picked up. He had no idea that his eyesight was any different from anybody else's—until the first hint that something was amiss came with the onset of night blindness. "I went to the opticians with my glasses and said, 'I don't know what it is, but I don't seem to be able to see as well at night and I get real glare from lights—which is a problem because I ride around on a motorbike.'" The optician merely told him that perhaps he needed tinted glasses—a "solution" that only made things worse.

Finally, in 1988, he was diagnosed with retinitis. His eyes continued to deteriorate—the tunnel of vision got smaller and smaller, and he also began to lose sharpness of vision. In spite of this (or perhaps to spite it), he was inspired by the intervention of an Olympic swimmer he was coaching to launch his own athletics career: "She went to the Seoul Olympics in 1988 and came back and told me about the Paralympics. She gave me her commemorative medal from Seoul and said, 'I want you to give me your commemorative medal from Barcelona'—the next Olympic venue. I laughed and said, 'I'm too old and too slow,' and she said, 'Well, you're only a year older than me, and we need to find out if you're too slow!' So that was it—once I'd made my mind up to do it I gave it 100 percent."

In 2011 Tim's son alerted him to a trial being conducted at the John Radcliffe Hospital in Oxford by Robert MacLaren. It involved the implant of an electronic retinal chip, and MacLaren was recruiting patients with end-stage photoreceptor loss—total night blindness. Tim wasn't attracted to the idea at first because he has always believed the best policy for someone with his impairment is to adjust, live, and compete in a disability world—to come to terms with it. And he didn't think the retinal chip would offer him any dramatic improvement: "I said, 'I don't really need to. I'm happy with my lot. It's not going to help me because it would have to be something really good to override how I function now. It would have to be a game-changer.'" Then his nerdy side took over, and he became fascinated by the technical details of how the device worked. Could his analytical mind bring something to the development of the chip that other patients' couldn't? He applied. MacLaren's team made it quite clear that, at worst, it might occlude what tiny residual light sensitivity he had left, and at best it could only give him a little improvement in light-efficient vision, to help him move around. He should expect nothing more than that.

The device was inserted during a ten-hour operation in October 2012. The surgeons first fashioned a small pit in his skull, behind the ear, and placed a power pack under the skin. They then threaded a cable along his temple, underneath the muscle, through a hole in his eye socket, and then painstakingly inserted the chip in a position in his retina where it was hoped it would replace lost photoreceptors. "When it was first switched on, it was one of those 'wow' moments, and I actually saw a light straightaway—a very bright light. It was the equivalent of being in a dark room and somebody suddenly strikes a match—it flares up." Tim finds that in a laboratory setting, he is able to read the white hands on the black face of a clock and can tell the correct time nine times out of ten. It helps if objects are stationary or if he is not moving. In everyday life, he is able to identify some objects around the office. For instance, he can tell if the computer screen is on and identify the outline of the screen but not what information is on it. In dark conditions outside, he can see streetlights and the illuminated pedestrian refuges in the middle of road crossings, but he still struggles in bright daylight.

Tim reckons that he could host his retinal implant for life, provided it doesn't malfunction and there is no sign of rejection of it by his eye tissue. But the importance of this research—and the collaboration of pioneering implant patients like Tim—is not necessarily because it tells us what advanced eye repair of the near future will look like. It is what the Oxford scientists have learned through it about the ability of cells in the retina to form new connections between photoreceptors and the optic nerve to the brain. Conventional wisdom had suggested that when there is any advanced degeneration in neural systems, as in spinal cord or brain injury, the nervous system couldn't repair itself. Once photoreceptors in the eye died off, did it mean that the other retinal cell types, like the bipolar cells and ganglion cells that relay visual information from the photoreceptors to the brain, would perish with them because they were receiving no inputs? The retinal implant, however, was able to forge new connections with output neurons to the optic nerve such that some degree of vision was restored. It showed conclusively that the rest of the neuronal visual apparatus was still intact. Specifically, the 1,500 light-sensing diodes on the chip stimulated these downstream neurons and allowed the brain to assemble a pixelated image. This led the Oxford scientists to believe that it might be possible to create a biological equivalent of the retinal chip. If they could manufacture a source of thousands of retinal photoreceptors—rods and cones—and implant them into the correct layer and with the correct orientation in the retina, they might be able to replace dead photoreceptors with new ones and bring back sight. This Oxford research is one of a number of initiatives, worldwide, that are learning how to manufacture and implant different types of precursor cells into the retina. Their insights stem directly from a growing understanding of the developmental biology of the vertebrate eye—the way this extraordinarily complex organ assembles in the embryo from a patch of undifferentiated cells to produce an astonishing visual device.

Of fundamental importance is the optic cup, the chalice-shaped chamber of the eye, which is made up of several tissue layers. To the outside of the eye lies the tough fibrous limiting membrane or sclera; inside that lies the blood-filled chorion. Then comes the retinal pigment epithelium, which supplies nutrients to the photoreceptors above it and takes away waste products. The photoreceptor layer is composed of over 120 million rods, which are particularly sensitive to low light levels and give us our night vision, and 6 or 7 million cones, which react to higher light levels and allow us to see in color. Rods predominate in the peripheral retina while cones predominate in a tiny yellowish patch only one and a half millimeters in diameter called the macula. Their concentration is highest in the fovea at the center of the macula, which is only 0.3 millimeters in diameter and yet is responsible for all our high-acuity central vision. This is the tunnel that was left in Tim Reddish's visual field after retinitis had killed off his peripheral vision.

In the vertebrate retina, the photoreceptors project their connections forward toward the center of the eye to a layer of bipolar cells, which in turn connect to an inner layer of ganglion cells. It is the axons, or projections, from these cells that all join together in the bundle of nerve fibers we know as the optic nerve. These inner layers of retina are also filled with Müller cells, amacrine cells (a form of neuron), and horizontal cells, all of which help relay signals from photoreceptors to the optic nerve. So the cells that actually absorb the photons lie to the outside of the retina such that light has to traverse all the other cell layers, and their connections, to get to them. Further, because the axons of the ganglion cells run across the inner surface of the retina before plunging back through it, en masse, as the optic nerve, we have a blind spot that contains no photoreceptors whatsoever. You can find this blind spot very easily by covering one eye and fixating on a point in front of you with the other. Then move a pencil horizontally to your right. Within a couple of inches it will briefly disappear before reemerging as you slowly track farther rightward. Slight movement up and down at the point of disappearance also causes the pencil to reappear in your field of vision.

On top of all this complexity in the retina we have the lens, which must form in the correct position to focus light upon the retina and is composed of a gradient of crystalline protein molecules to give it its variation in refractive index; and an iris, which changes the size of the pupil and controls the amount of light that falls on the retina. Not to mention the musculature that controls the movement of the eye.

This has barely scratched the surface of the eye's structural complexity, but it has probably done enough to help us realize why the eye, above all complex organs in the body, has served as a battleground for the warring armies of creationism and evolution ever since the early nineteenth century. And despite 150 years of refinement of Darwinian evolutionary theory, creationist dissent is, today, as loud as it ever was.

The most well-known historical proponent of intelligent design was the Reverend William Paley, a moral philosopher who became Archdeacon of Carlisle, in the north of England, in the late eighteenth century. In his famous treatise *Natural Theology*, he laid out his argument for the existence of God, and for God's agency in the design of nature. He famously pointed out that if he dislodged a stone while walking in the countryside, there could be no objection to his assumption that it had hitherto lain there, inert, for all time. It had no purpose, as such, and required no further explanation. Whereas if he discovered a watch lying on the ground, he would have to conclude, when inspection revealed the many parts of its complicated internal mechanism, that those parts betrayed the fact that it had a purpose. It had a design. The watch had to have had a watchmaker who intended it to keep time. Clearly one only had to look at the construction of any animal species to conclude that in its complexity it also had been designed for a purpose. But since no animal could comprehend its own construction or understand its use by virtue of its own design, the author of that purposeful design must have been God. God was the "watchmaker" of all living things.

Half a century later, along came Charles Darwin, to argue the opposite—that animals and their organs, even something as spectacularly complicated as the vertebrate eye, arose through natural selection. All that was needed, he argued, was proof of numerous stepwise gradations from some primitive structure to a fully evolved eye, each of which had measurably improved the survival of its owner to the extent that it would have been selected for. However, when Darwin published *The Origin of Species* in 1859, he was unable to supply a chain of intermediate forms as proof of his theory. Not surprisingly, creationists are rather fond of the following quote: "To suppose that the eye, with all its inimitable contrivances . . . could have been formed by natural selection, seems, I freely confess, absurd in the highest possible degree. . . ." Though they are rather less interested in the rest of Darwin's passage in which he begins to build his argument: "Yet reason tells me, that if numerous gradations from a perfect and complex eye to one very imperfect and simple, each grade being useful to its possessor, can be shown to exist; if further, the eye does vary ever so slightly, and the variations be inherited, which is certainly the case; and if any variation or modification in the organ be ever useful to an animal under changing conditions of life, then the difficulty of believing that a perfect and complex eye could be formed by natural selection, though insuperable by our imagination, can hardly be considered real."

Darwin then went on to suggest a highly plausible route by which a fully functioning vertebrate eye could evolve in gentle steps from a primitive origin in a single photosensitive cell on the skin. This proceeded via a clump of such cells, to pigment cells forming a slight depression that became deeper, to an optic nerve connecting to the pigment cells with a translucent layer of skin growing over them, to a lens subsequently forming from that translucent layer.

Creation-minded commentators ever since have preferred to view the problem in a different way. Unimpressed by Darwin's gradualism, they see irreducible complexity by which the eye is the sum of a number of hopelessly complex parts, the absence of any one of which would render the whole thing useless. This quote from Francis Hitching's 1982 book *The Neck of the Giraffe* is now a little dated but remains typical of the creationists' arguments concerning the origin of the eye:

> For the eye to work . . . it must be clean and moist, maintained in this state by the interaction of the tear gland and moveable eyelids. The light then passes through a small transparent section of the protective outer coating, the cornea,

> and continues via a lens that focuses it on the back of the retina. Here 130 million light-sensitive rods and cones cause photochemical reactions that transform the light into electrical impulses that are transmitted to the brain for appropriate action. Now it is quite evident that if the slightest thing goes wrong en route—if the cornea is fuzzy, or the pupils fails to dilate, or the lens becomes opaque, or the focusing goes wrong—then a recognizable image is not formed. The eye either functions as a whole or not at all.

Darwin's present-day bulldog is, of course, Richard Dawkins, and he savages the above quote in his 1986 book, *The Blind Watchmaker*. As Dawkins pointed out then, and as Tim Reddish and millions of sufferers of eye disease will tell you today, a very rudimentary eye or a human eye hopelessly compromised by disease is still much better than no eye at all. *The Blind Watchmaker* is Dawkins's riposte to Paley, Hitchings, and any other peddler of the concept of irreducible complexity. In 1987 I teamed up with Dawkins to make a documentary film based on *The Blind Watchmaker* for BBC Television. We explained the evolution of the eye in an animated graphic sequence that showed how a simple row of photosensitive cells on a skin surface could register the presence or absence of light by being switched either on or off. We then dropped the cells into an ever-deepening pit until they had formed a pinhole camera capable of discriminating objects because light from them fell only on discrete patches of "retina." We then imagined that the cells happened to secrete mucus that, over time, gathered into a ball and blocked the pinhole, forming a crude lens. It was clear that this would allow much finer discrimination by focusing incoming light onto the photoreceptor surface. Dawkins and I thought the sequence very convincing, but I was surprised to discover that many of my fellow producers were far from bowled over and considered it simply a cartoon. We hadn't "proved" anything. Well, of course, in the absence of any underlying mathematics, we hadn't produced a scientific proof of a Darwinian process—merely a plausible scenario—but two scientists from Sweden have constructed an extraordinary mathematical model of eye evolution that shows that eyes *can* evolve in exactly the way that Darwin, and Dawkins after him, suggests, and in a remarkably short period of time.

In 1994 Dan-Eric Nilsson and Susanne Pelger, from Lund University in Sweden, published what they called "a pessimistic estimate of the time required for an eye to evolve." In their model, evolutionary selection works on one single parameter of the eye—spatial resolution, otherwise known as visual acuity. Any incremental change in the structure of their model eye that increased the amount of spatial information the eye could detect would be selected for. It is important to stress that theirs was a mathematical model with real, conservative values for morphological variation, intensity of selection, and heritability.

Their starting point was a circular patch of light-sensitive cells sandwiched between a transparent protective layer and a layer of dark pigment. They then subjected this structure to selection pressure favoring an increase in the ability to distinguish fine detail. In the first stage, the photo-sensitive layer and the pigment layer folded inward to form a pit while the transparent layer deepened to form a gel-like vitreous body that gave the proto-eye the shape of a hemisphere. This improved vision up to the point at which the depth of the pit equaled its diameter, and things only improved further if the circular lip of the pit then constricted to narrow the aperture down to a pinhole. The eye by this stage resembled a circular hollow ball with a narrow circular opening at the top—rather like a round vase. However, as the aperture constricted and the image became better resolved, it also became "noisier" because the small numbers of photons being captured became drowned out by random events that distorted fidelity. At this point no further improvement in acuity could occur without the introduction of a lens. The continued selection pressure caused the gel in the center of the eye to form a crude elliptical lens that first plugged the pinhole and then, under further selection, moved to the center of the eye, decreasing slightly in diameter, becoming rounder, and increasing in refractive index (its ability to bend light) as it went. A flat iris then formed to cover lens and pinhole.

The end point was a model aquatic eye extremely similar to that of a squid or octopus. They calculated the number of 1 percent changes in spatial resolution that occurred in each of these stages and came up with a total of 1,829 steps of 1 percent. How long would the evolution of their eye have taken in real time? They fed in their conservative parameters for heritability, strength of selection, and variation in each of these components of the eye, assumed a generation time of one year (about average for small to medium-size aquatic animals), and calculated that it would take less than 360,000 years for an aquatic camera eye to evolve from a flat patch of light-sensitive cells. This is lightning speed in evolutionary time. The first recorded eyes come from the early Cambrian period some 550 *million* years ago. Enough time has elapsed since then, they contend, for eyes to have separately evolved more than 1,500 times!

Of course, this does not mean that evolution *did* reinvent the eye 1,500 times during the evolution of life on Earth. It only shows that an optically good eye *can* evolve by natural selection in an astonishingly short period of time compared with the huge span of geological time over which we know that eyes have existed. And, gratifyingly, Nilsson and Pelger's model eye corresponds very well with the enormous diversity of eye structure to be found in nature: "Eyes closely resembling every part of our model sequence can be found among animals existing today. [For instance,] from comparative anatomy it is known that mollusks and annelid worms display a complete series of eye designs, from simple epidermal aggregations of photoreceptors to large and well-developed camera eyes."

So, how many times has an eye independently evolved during the history of life on Earth? No one really knows for sure. The great evolutionist Ernst Mayr, together with Luitfried von Salvini-Plawen, estimated, from the enormous anatomical variation among eyes and the photoreceptors within them, that they may have separately arisen between forty and sixty-five times. Walter Gehring of the University of Basel in Switzerland, by studying the evolutionary history of genes that have controlled eye formation throughout the animal kingdom, came to a very different conclusion.

Gehring focused on an ancient gene named Pax6, which has been at the top of the hierarchy of all genes involved in making eyes since the time of the oldest ancestors of all vertebrates and invertebrates, collectively known as the Bilateria. They were the first primitive animals to have bilateral symmetry. In fact, close relatives of Pax6 have been found in species that even predate the Bilateria, such as the box jellyfish, sponges, and the polyp *Hydra*, which people of my generation, at least, will remember investigating down the microscope in biology class.

Gehring proposed that Darwin's prototypical eye, comprising only a photoreceptor and a pigment cell, would be controlled by Pax6 and that all the eyes in the animal kingdom have evolved by gradual stepwise insertion of subsidiary genes to form a hierarchy with Pax6 at the top. The breathtaking, and unavoidable, conclusion from his synthesis is that eyes only evolved once in the animal kingdom 700 million years ago, and all the eyes we see today, or have ever existed, are stepwise evolutionary developments in complexity controlled by stepwise additions to this collection of genes, so that a complex human or insect eye today is controlled by a family of some two thousand genes.

Gehring's insight that Pax6 might be ubiquitous in eye formation came from the work of other scientists, which showed that the "eyeless" mutation in fruit flies, the "small eye" mutation in mice, and a hereditary syndrome called aniridia in fetal humans, where fetuses abort because they lack eyes and noses and have substantial brain damage, all correspond to Pax6. A series of ingenious experiments introduced the mouse mutation into developing flies to produce a range of ectopic eyes—eyes developing where they shouldn't—on antennae, wings, and legs. This showed that Pax6 promotes eye formation whenever it is turned on, wherever it is. Further experiments suggested that Pax6 not only started off the process of eye development but also initiated the further differentiation of cell types in the eye. As Gehring points out: "These experiments lead to the conclusion that Pax6 is a master control gene on the top of the genetic cascade leading to eye morphogenesis and that this master switch can initiate eye development in both insects and mammals."

Darwin's prototype actually exists, in the form of the very rudimentary multiple eyes of planarian flatworms, which only consist of one photoreceptor and one pigment cell. Since near-identical versions of Pax6 are found in planarians, and all the way up to the human eye, Gehring concludes: "All bilaterian eye types go back to a single root, a Darwinian prototype as found in planarians. Starting from this prototype, selection has generated increasingly better-performing eyes and these various eye types arose by divergent, parallel and convergent evolution."

Most evolutionary biologists today believe that, in his sweeping scenario, Gehring has overstated his case. Although important, Pax6 is not quite the "master-controller" of eye formation that Gehring proposed. There are too many instances, throughout vertebrates and even fruit flies, where the first stages of eye development still occur when the Pax6 gene is knocked out. And eyes in flatworms can form completely in the absence of Pax6. Pax6, although very important, is only one of an orchestra of genes that steer eye development. The presence of an ancestral member of the Pax gene family in an ancient animal is no guarantee that it was actually associated with eyes.

All evolutionists agree, however, that the main evolutionary spadework creating the plethora of eyes we see today occurred, as the zoologist Andrew Parker pithily puts it, "in the blink of an eye" during the Cambrian explosion of forms approximately 530 million years ago. The earliest representatives of all the major animal taxonomic groups (phyla) alive on Earth today originated within a few million years. Here, the trilobites—ancestral arthropods (an animal phylum whose present-day members include insects, spiders, and crustacea)—rapidly evolved compound eyes, while primitive vertebrates like *Haikouichthys*, thought to be the ancestor of fish, contained two sophisticated camera eyes, complete with a lens, in a tiny aquatic animal less than an inch long and weighing less than one ounce. Almost every type of eye we find today—from a simple pinhole, to a compound insect-type eye, to the sophisticated camera eye of vertebrates and some mollusks—evolved at this time. It is thought that the fuel for this high-octane evolutionary sprint was visually guided predation. Sharp-eyed predators provided the selection pressure for sharper eyes among their prey, to avoid being eaten, which, in turn, selected for better visual acuity in predator species—and so on.

Is it possible to go back in time before the Cambrian explosion—did even more ancient animal species have eyes? The question is difficult to answer because such tiny, soft-bodied species leave little or no trace in the fossil record. But we do know that the ancestors of all animals—the Bilateria—existed between 300 and 100 million years before the Cambrian period, and the existence of the major building blocks for vision—specialized cells called photoreceptors and the light-sensitive molecules they contain—can be traced back to that time. As Nilsson points out, photoreceptor cells are not exclusive to eyes and exist in a wider sensory context in plants, fungi, and many unicellular organisms. The trick, for early animals, says Nilsson, was to press photoreceptors into service to provide eyes and thus release enormous potential for mobile organisms. So, photoreceptors evolved before eyes did. The vast majority of animal photoreceptors fall into two distinct types: rhabdomeric and ciliary. In the former, the visual pigment molecules are arranged on specialized outgrowths of cells called microvilli; on the latter, they are arranged on folded membranes of long, slender whip-like organelles called cilia. The visual pigments are different between the two forms, as are key chemicals in the cascade that turns light into nerve impulses. Today most invertebrates have rhabdomeric photoreceptors, and vertebrates have ciliary photoreceptors. According to Nilsson, the differences between the two forms of photoreceptor are profound and point to a separate evolutionary origin. It is possible, explains Nilsson, that the common ancestor of all bilaterian animals had both types of photoreceptor, whether or not they were actually used for vision. Invertebrate ancestors split off from this bilaterian stock, taking rhabdomeric photoreceptors with them, and vertebrate ancestors incorporated ciliary receptors in their visual systems. Thus, although eyes have independently evolved a number of times over the history of life on Earth, from different embryological substrates and over a wide range of sophistication, they have always done so by selecting from this ancient eye-building kit of photoreceptors and their pigments.

In a wonderful piece of molecular paleo-archaeology, Davide Pisani, from the National University of Ireland, together with colleagues from the University of Bristol, has focused on the family of light-sensitive pigments found in photoreceptor cells that transforms photons of light into electrochemical signals. They are collectively known as opsins. Pisani and his colleagues produced a taxonomic tree of the different opsin visual pigment molecules that leads right back beyond the Neuralia, the umbrella animal group that includes the cnidarians (sponges and jellyfish), ctenophores (comb jellies, sea gooseberries, sea walnuts, and Venus's girdles), and bilaterians (the ancestors of all other animals on Earth) to originate in a strange group of incredibly simple amoeba-like animals called Placozoa. It was here, they theorize, that the fundamental ancestor molecule of all opsin photopigments underwent a gene duplication to produce an ancestral opsin gene and its sister, the gene for melatonin, the chemical that is now involved in the control of circadian rhythms. This opsin then underwent a period of rapid-fire evolution over a short period of 11 million years to produce opsins that could detect light. Pisani's work places the origin of the opsin genes at approximately 700 million years ago and lays the foundation for studies to further refine our knowledge of the evolution of the opsin gene family right up to the vertebrates and the evolution of color vision.

When the environment presents similar levels of selection for acute vision—perhaps in order to improve the success of predation or to avoid it—animal eyes that have very different evolutionary origins can converge on very similar kinds of sophisticated eyes. Vertebrate and cephalopod (squid and octopus) eyes are a classic example of how convergent evolution has produced camera eyes that, superficially at least, are very similar. Both eyes have a cup-like chamber, both have sophisticated retinas densely packed with photoreceptors, and both have a lens. Yet cephalopods are mollusks and evolutionarily remote from vertebrates.

Ironically, this extraordinary example of convergent evolution has produced a battleground for the warring armies of creationism and evolution. Creationists, as we have seen, follow a form of argument first put forward by William Paley, that the eye, with all its complex interrelating parts, could not have been produced by stepwise evolution. Evolutionists have found themselves countering this argument by inviting invidious comparison between the aesthetically pleasing optics of the squid eye and the rather messier construction of the vertebrate, hence human, eye.

The main difference that concerns evolutionists is the orientation of photoreceptors in the retina, which is the inevitable consequence of the different embryological origins of eyes in vertebrates and cephalopods. In the vertebrate eye, the retina develops from the central nervous system and arises as a bud, or swelling, of the neural tube that will eventually form the frontal part of the brain. This optic vesicle rises to kiss the outer skin of the embryo, stimulating it to develop a lens. It then invaginates to form the eye-cup. This results in the so-called inverse or inverted retina, where the photoreceptors actually point away from the light while the connections that lead from them to the optic nerve face inward toward the lens. In invertebrates the whole eye develops from the skin by a series of complex tissue foldings that result in the photoreceptors facing toward the light and their connections to the optic nerve and the brain facing backward toward the rear of the eye. This means that their reception of light is unimpeded by overlying cell layers and the jumble of nerve connections to the brain. Furthermore, because the neurons run from the back of the photoreceptors to the optic nerve in octopus and squid, they do not have to cross the inner surface of the retina before plunging through to the back of the eye en route to the brain. Squid and octopus eyes, consequently, have no blind spot. This is the everse or everted retina, and, at first glance it appears that in the squid and octopus, evolution has achieved a cleaner and more pleasing design for the eye than the untidy jumble—complete with blind spot—that is the vertebrate eye.

Consequently, when they argue with creationists, even august professors of evolutionary biology can get themselves into a tangle. Part of the reason is that, in order to convince us of the hand of evolution in the "design" of the human body, rather than the hand of God, they often resort to what is known as the "argument from poor design," and this argument is frequently deployed when defending the evolutionary origin, as opposed to the putative celestial origin, of the human eye. The "design flaw" invariably singled out is the inverse retina. This is how Richard Dawkins sees the engineering fiasco of the vertebrate eye in *The Blind Watchmaker*:

> Any engineer would naturally assume that the photocells would point towards the light, with their wires leading backwards towards the brain. He would laugh at any suggestion that the photocells might point away from the light, with their

wires departing on the side *nearest* the light. Yet this is exactly what happens in all vertebrate retinas. The wire has to travel over the surface of the retina, to a point where it dives through a hole in the retina called the blind spot to join the optic nerve. This means that the light, instead of being granted an unrestricted passage to the photocells, has to pass through a forest of inter-connecting wires, presumably suffering at least some attenuation and distortion—actually probably not much, but, still, it is the *principle* of the thing that would offend any tidy-minded engineer!

The joint father of the field of evolutionary medicine, Randolph Nesse, employs a very similar argument. In a well-trammeled YouTube conversation with Dawkins, he exclaims: "Imagine a camera designer from Nikon or Pentax who put the wires between the light and the film—which is how our eye works. And not only that—our eye has a whole blind spot where nothing works at all!" The inverse retina, claims Nesse, also leaves us susceptible to detached retinas. Dawkins replies: "Helmholtz, the famous German psychologist, once said that if an engineer had given him the human eye, he'd have sent it back!"

According to Nesse, the human eye is no match for the eye of a squid or an octopus. "All cephalopods have an eye that is designed properly. Their eyes have all of the vessels and nerves coming right through the back of the eyeball so that they can't get retinal detachment. They never have a problem with the blind spot, so they don't need to move their eyes as much as we do to get a complete field of vision. It's a better design, entirely, than ours!"

To listen to Dawkins and Nesse is to get a picture of a vertebrate eye riddled with almost absurd design flaws. They are, in effect, saying: "By flawed design you can be sure you are looking at the hand of evolution. God, like any competent engineer, would never have made such mistakes!" I'm uneasy about this argument from poor design because, unwittingly or otherwise, the evolutionists who employ it invite us to see evolution as a hopeless bungler forever doing botched jobs like a cowboy builder. I think evolution is better than that, and I think Dawkins and Nesse may be missing a trick by not exploring further what might be the very positive, or adaptive, aspects of the inverse design of the vertebrate retina. After all, vertebrates, all with inverse retinas, have become wildly successful animals. So do vertebrate eyes really compare dismally to cephalopod eyes?

Several scientists have looked for the plus side of vertebrate retinal design, and their theories have turned Nesse's and Dawkins's arguments inside out. Ronald Kröger is a professor in the Vision Group at Lund University. "The real question," he says, "is why animals with inverse retinas have been successful and radiated into a group, the vertebrates, with the most well-developed visual systems." The reason for the success of animals with inverse retinas, to him, is both obvious and straightforward: the inverse retina is actually the superior solution. "It has so many advantages that I am actually much more surprised by the fact that other animal groups, such as cephalopods, have evolved similar eyes with everse retinas!"

The main advantage, Kröger believes, together with colleague Oliver Biehlmaier, lies in saving space. Kröger is an expert on the eyes of small fish, which need sharp eyesight but only have small bodies. In their eyes the space between the lens and the retina, normally filled by vitreous humor in vertebrates, is almost totally taken up with the retinal cells that process the visual information passed to them by the photoreceptors. If all these cells were placed outside the photoreceptor layer, as in an everse design, the eye would have to be much larger to accommodate them. Kröger and Biehlmaier calculate, for instance, that the smallest eye that could accommodate an inverted retina 100 micrometers thick is 330 micrometers in diameter with a lens 130 micrometers in diameter. With the same size and focal length of lens, an eye with an everted retina would have to be 420 micrometers in outer diameter and twice the volume. So, if you want a sophisticated retina but don't want a goggle-eye, go for an inverted retina. Since the ancestors of most animal groups tended to be very small, the adoption of an inverse retina could have been crucial. Many animal larvae are also extremely small but need good vision to survive. Of course, as animals become bigger, the space-saving property of an inverted retina becomes less dramatic. In a modeled simplified eye the same size as the human eye, for instance, Kröger and Biehlmaier calculate that the inverted retina results in an eye volume 11.3 percent less than would be the case were the retina everted, and they guesstimate a space saving of approximately 5 percent for the actual human eye—less dramatic but still a considerable saving in evolutionary terms. In addition to the space-saving advantage, they contend, the inverted retina brings the photoreceptor outer segments in close proximity to the retinal pigment epithelium that regenerates their visual pigment. It also allows for the nourishment of the metabolically greedy photoreceptors via the choroid blood supply, while keeping light-absorbing hemoglobin out of the path of incoming light.

The most important property of the inverted retina is increased thickness, which translates into more powerful and complex retinal processing. This means that more processing of visual information can be done in the eye before signals are sent down the bottleneck of the optic nerve into the brain. Neuroscientists Tim Gollisch and Markus Meister have pointed out that the vertebrate retina actually contains fifty clearly distinct cell types, far more than would be necessary for basic visual tasks like light adaptation and image sharpening. The multi-layered vertebrate retina is equivalent to a sophisticated laptop computer tethered to a mainframe desktop computer via an Ethernet connection. Much of the computation involved in detecting rapid movement and making sense of the myriad features in the field of vision sampled by rapid eye saccades is done in the vertebrate retina, not in the brain. All this short-range computation involves cross-talk between retinal neurons and is analog rather than digital; it uses graded potentials rather than digital on/off signals. This allows, says Kröger, for enormously higher information densities and is a computer designer's dream come true. The final output to the brain from the ganglion cells is converted to digital because the relatively long communication distance to the brain would otherwise make the signals more susceptible to interference.

There are over 100 million photoreceptors in the human retina but only 1 million axons in the optic nerve. If every photoreceptor were wired independently to the brain, the optic nerve would have to be wider than the diameter of the eye. This is why the vertebrate "smart" retina, which achieves a great deal of local pre-digestion of visual information before passing signals on to the brain, is particularly adaptive. It is even more important in lower vertebrates, whose brains are far smaller than the mammalian brain but for whom high-acuity vision is just as important for survival. For instance, Kröger would not be surprised if the number of neurons in the eye of a crocodile exceeds the number of neurons in its brain! Birds are a particularly good example of the advantage of an inverted "smart" retina. They need to save space and weight in all departments in order to get airborne and fly with agility and endurance. Most bird species also have extremely sharp vision. Their retinas are very thick to allow for even more advanced signal processing before visual information is sent to their relatively small brains.

It is absolutely true that in the strict sense of an optical device like a camera, the octopus eye is superior to the vertebrate eye, because there is no cellular jumble between light and the photoreceptors. But octopuses can only see in black-and-white—they have no color vision. And, as we have seen, the vertebrate retina is not just a device for photon capture but a formidable signal processor. In contrast, cephalopods have none of the complex multi-layered structure typical of the inverse vertebrate retina. They are forced to send raw visual information for further processing to two structures called the optic lobes, which are outgrowths of the brain that lie behind the orbit of the eye. Each and every photoreceptor is wired independently to the optic lobes of the brain, which results in many relatively long-distance connections that can pick up noise on the way. This inefficient wiring, Kröger points out, also costs time and makes the fast comparison of inputs more difficult between neighboring photoreceptors, so important for the computation needed for color vision, movement detection, and the quick picking out of features in the field of vision. He concludes: "Vertebrates have evolved into the group of animals which most heavily rely on vision with high spatial resolution. The inverted retina has most likely been an important factor since it allows for massive retinal processing of visual information without investment of precious space and weight."

In an amusing postscript to this inverse/everse retina debate, there was great excitement in creationist circles in 2007 caused by the publication of a paper by Kristian Franze and colleagues. Creationist literature has regularly featured their work ever since because it seems to show that God knew what he was doing all along in the creation of the vertebrate retina. Franze looked at Müller cells—long, thin, tubular cells that span the entire retina and appear funnel-shaped at each end. These cells were thought to be one of the many support cells in the retina, but Franze claimed that they were orientated along the line of light propagation onto the retina, and advanced microscopy techniques, he reported, suggested that they were able to provide a low-scattering passage for light from the retinal surface to the photoreceptor cells. Using lasers, they determined that the Müller cells acted just like the optical fibers—waveguides—that carry broadband and television pictures into many of our homes: they were nature's own fiber optics. By conducting low-distortion images through pipes from the retinal inner surface to the rods and cones, they seemed able to bypass all the jumble of blood vessels, retinal cells, and axons that lie between the photoreceptors and incoming light, and would otherwise somewhat degrade the visual image. Franze's discovery was much trumpeted by creationists because it appeared to show that the inverse retina was, after all, the cunning design of an intentional engineering genius and not the poor design of blind evolution. It put evolutionists on the back foot; they were forced to argue that the Müller cells were typical of evolution's tinkering by applying a retro fix to a flawed design. They need not have bothered tying themselves in knots.

It appears that much of the ophthalmology research community fails to share the creationists' enthusiasm for Franze's fiber optics. Although they have received no wide attention whatsoever, there are fundamental and devastating criticisms of Franze's work. David Williams is director of the Center for Visual Science at the University of Rochester, New York, and a leading expert on human vision. There is, he says, unequivocal evidence that Müller cells are not waveguides in any meaningful sense of the word. The anatomy of the retina reveals their irrelevance because it turns out that the Müller cells do not, after all, point toward the light streaming in through the pupil. In the fovea, the small high-acuity vision pit in the retina, the Müller cells are splayed apart, and throughout the peripheral retina, Williams explains, Müller cells are orientated radially, that is toward the center of the eye, not toward the pupil. It is the *cones* in the peripheral retina that are the proper waveguides because they do what they should — they point toward the light. We have the ability to see rod and cone photoreceptors in the retina using adaptive optics, he points out, precisely because they are waveguides. They direct light that enters them from a light source straight back through the pupil into a camera, where they appear as bright spots. Müller cells never light up in this way.

In April 2011 the journal *Nature* published a paper containing a link to a beautiful movie that showed an eye-like structure emerging through time. This movie had not been taken to show the formation of the eye in a living embryo; the eye that assembles itself quite literally before our eyes was grown in vitro. It is an eye in a dish. This dramatic and exciting experiment's importance cannot be overestimated because it has told us a huge amount about how eyes develop and has shown us that the blueprint for an eye is actually contained within the differentiating cells of the eye itself: in other words, eyes self-organize — there is no outside prodding. And this experiment has provided proof-of-concept and inspiration to a community of ophthalmic scientists all over the world who are using the principles of developmental biology to regenerate diseased eyes.

Mototsugu Eiraku, the late Yoshiki Sasai, and their colleagues from the RIKEN research institute in Japan started by placing a blob of mouse embryonic stem cells in a dish together with some culture medium and a gel-like substrate called Matrigel, to give the cells a membrane to hold them together. They then gave the blob a few doses of a protein called Nodal, which is known to be important for differentiating tissues. Six days into the experiment, a number of hollow spheres formed that rapidly developed into hemispherical sacs or vesicles. The scientists could watch it happening because they had introduced the gene for green fluorescent protein (GFP) into the stem cells. (GFP was originally isolated from a jellyfish and is widely used as a marker for visualizing tissue development.) When the cells began to differentiate, they switched GFP on and the developing structure glowed a ghostly green. Thus they had created their own video graphic! Without the GFP, the vesicles would have developed unnoticed until enveloped in surrounding tissue.

Between days eight and ten, the vesicle dramatically changed its shape, folding inward to form a cup exactly the same size as the optic cup in embryonic mice. The cells of the outer part of the wall of the cup began to extrude the protein marker molecules you would expect to see in developing retinal pigment epithelium, while the cells on the inner side of the wall expressed markers synonymous with retinal neurons. Accepted wisdom says that a lens, developed from ectodermal tissue, is necessary to induce the formation of an optic cup, but Eiraku and Sasai's cup developed completely under its own steam with no outside interference. They then carefully excised several of the vesicles and grew them on, isolated from the original stem cell aggregate. Amazingly, within fourteen days, their "embryonic retina" started to differentiate all the different known cell types: photoreceptors, ganglion cells, bipolar cells, horizontal cells, amacrine cells, and Müller glial cells. Furthermore, these different cell types arranged themselves in the correct anatomical order you would expect to see in a neonatal eye, with the bipolar cells overlying the photoreceptors and the innermost layer of retina being composed of ganglion and amacrine cells. They had built an optic cup and retina, or, rather, these structures had built themselves: they had become a DIY eye. As Eiraku and Sasai conclude, "This complex morphogenesis possesses a latent intrinsic order involving dynamic self-patterning and self-information driven by a sequential combination of local rules and internal forces within the epithelium." They looked toward a "just around the corner" future for regenerative eye medicine in which DIY-eye technology will be able to grow fully stratified 3-D neural retinas to order, in sheets.

Of course, the more curmudgeonly reader might quite rightly point out that the Japanese scientists have not built an eye from scratch—only a retina and optic cup. What about a lens, for instance? Fortunately, Andrea

Streit from King's College London has demonstrated exactly how the developing optic cup of the eye induces the formation of a lens directly over it. The outer cell layer of any embryo—the ectoderm or skin—has an intrinsic ability to form a lens at any part of its surface. This is why experimenters have been able to induce lenses all over the developing body of a frog or an insect, for instance. Streit has shown how a wandering population of neural crest cells that lies between the developing central nervous system and the skin normally inhibits the ectoderm from lens production. They operate cell-signaling pathways that inhibit that important gene for eye formation, Pax6. However, when the optic cup forms and rises up to kiss the overlying skin, it forms a cordon sanitaire that isolates a local area of skin from which neural crest cells are excluded. The inhibition of Pax6 is lifted, and a lens forms, just where it ought to be.

Robin Ali, from the Institute of Ophthalmology in London, is excited by the Japanese research. He moved into stem cell eye regeneration in 2003 and has been steadily gearing up to human trials for a decade. Ten years ago, stem cell science was a hubristic world full of wild claims, he says, backed up by poor-quality data. What was needed was painstaking research with animal models to see precisely what was possible, and what was not. He uses mice because the mouse retina is still developing after birth. "I wanted to know whether it was possible to transplant a photoreceptor cell at all—never mind a stem cell—is it technically possible? Will it engage with the host retina?" He began by taking photoreceptor cells from a three-day-old mouse and transplanting them into the retina of another mouse at the same stage of development. Sure enough, they integrated perfectly. "That told us what a transplanted photoreceptor cell should look like, whether from a stem cell or whatever—integrated, right way up—not a blob, not a mess, not an artifact."

Ali's team then compared transplantation success with retinal stem cells ranging from three days to three weeks old to find the optimum age for transplant. "We could look to see the efficiency of integration of those cells. It followed a bell curve. If you took retinal stem cells, which are completely immature, they wouldn't integrate—they just developed into little retinas within the sub-retinal space where we'd injected them. They were true retinal stem cells, but they didn't want to know the neighbors. And if we transplanted fully mature retinal cells, they didn't do anything at all. So, at the height of the bell curve we found this window that corresponded with the birth of photoreceptor cells, whose peak is around five days after birth. That is the optimal age for the donor cell." The photoreceptor cells that transplanted most successfully were not stem cells but cells developed from them that had just ceased dividing but were still immature.

It took them another five years to prove that they could inject forty thousand of these rod precursor cells into a blind mouse such that they integrated and made functional connections within the retina, joined synapses with the bipolar cells, and were able to relay information to the brain. Experiments where they allowed the implanted mice to navigate their way through a maze showed that they really did have improved eyesight. It was at this point that they borrowed a technique from one of Eiraku's colleagues in Japan to turn embryonic stem cells into a variety of retinal cell precursors. Ali explains: "Since then we've adapted the Japanese protocol and so we have these self-organizing retinas from the embryonic stem cells. That's why I say that Eiraku's work was a landmark for regenerative medicine for the eye because this was making a retina in a dish that was equivalent to taking a neonatal mouse retina as a source." The beauty for Ali is that Eiraku's work cuts out the necessity for complicated cell culture and the headache of synchronizing the development of the precursor photoreceptor cells to make a homogenous suspension of cells all at the same, optimum, stage for implantation. Eiraku's retina has done all that for him. They now have the equivalent of a donor retina and have already transplanted it into mice.

Age-related macular degeneration destroys the photoreceptors in the central high-acuity part of the retina, the macula. It causes gradual loss of central vision, which first becomes blurred and then completely blind as the disease progresses. Jan Provis, professor of anatomy at the Australian National University and an expert on the retina, has a theory for the cause of macular degeneration that is built around the idea of evolutionary compromise; in this case, the trade-off of acute eyesight when we are young for visual impairment in later age. It is a classic case of "live now, pay later." The macula, she reminds us, occupies less than 4 percent of total retinal area, yet it is responsible for all our useful vision in bright light. It is made up of three concentric rings—the fovea, parafovea, and perifovea. The tiny central region, the fovea, contains the highest concentration of cones in the entire retina, together with some rods. It is estimated that a tiny lesion in the fovea could knock out 225,000 cones together with 25 percent of the ganglion cell output to the brain and render you legally blind.

Evolution has jam-packed the fovea with cones to give us acute eyesight, but that very high cone density has been achieved at the expense of adequate blood supply, even though photoreceptors are greedy and consume more oxygen than any other cell type in the body. They draw their blood supply from retinal blood vessels and from the chorion, the outer layer of the eye between the retinal pigment epithelium and the sclera, which drains into a network of blood capillaries called the choriocapillaris. In most parts of the body, blood supply to organs is regulated by the autonomic nervous system, which can kick in to increase blood supply to hardworking tissues. However, in the retina, there is no autonomic control to increase supply on demand. Worse still, in the foveola, the pinpoint center of the fovea where cone density peaks, there are no retinal blood vessels whatsoever! Adequate blood supply has been sacrificed to photoreceptor number. If choroidal blood supply is reduced for any reason, this will tip the whole central retina into borderline oxygen starvation. As Provis puts it: "The foveal region should be understood as an environment in which neurons are in critical balance with their blood supply—a balance easily disturbed by changes in blood flow, oxygen and nutrient delivery."

Blood supply to the central retina is in this critical condition all the way from embryonic development of the human eye to birth and adulthood because the cones differentiate in the embryonic retina before the retinal capillaries are fully established. Evolution has ameliorated the situation by thinning the retina in the foveal region, the so-called foveal depression. The bore of the capillaries has been increased, and Bruch's membrane, which separates the chorion from the retinal pigment epithelium, has been reduced in thickness. In young, healthy individuals, says Provis, this marginally increases oxygen and nutrient supply to the photoreceptors, but, ironically, it sets the scene for macular degeneration as we age because, over time, blood-borne waste products are more easily forced out of the choriocapillaris under hydrostatic pressure and accumulate in Bruch's membrane and the space immediately under the retinal pigment epithelium. These deposits contain lipids and are called drusen. Slowly they compromise blood supply to the retina, cause low-grade local vascular disease, and become inflamed. Measurements of fats in Bruch's membrane at the macula are seven times higher than in the peripheral retina. As the macula becomes more stressed, it secretes a cell-signaling protein called vascular endothelial growth factor that stimulates the production of new blood vessels. However, they tend to be weak and thin-walled and very prone to leakage. This is end-stage wet macular degeneration. There is little an ophthalmologist can do to arrest it besides using laser treatment to try to cauterize these rogue blood vessels.

David Lee suffers from a form of juvenile onset macular degeneration called Stargardt disease. His problems began to arise in 1988, when he was twenty-two years old: "My wife was learning to drive and her driving test was due so I asked her to read some number plates out for me as we were walking home. She read them as clear as day and I said: 'I can't see that!'" Thinking he must need glasses, he made an appointment with an optician, who shocked him by telling him he had to go to the hospital immediately. "He said, 'There's something dark appearing at the back of your eyes.' Within a fortnight I'd been diagnosed with Stargardt's, and they said there was nothing they could do."

At the time, David was the manager of a brick works, which involved a lot of temperature readings and digital read-outs at the kilns. They gradually got beyond him, and the company reluctantly had to let him go for his own safety. He and his wife bought a pie and sandwich shop that he still runs today, even though his eyesight is heavily compromised. While the shop girls run the counter, he bakes all his own pies and cooks the breakfasts in the back kitchen. He can even fill the sandwiches provided that all the ingredients are lined up close by in the right order. At nighttime he finds the glare of car headlights very painful. He tends to stick to familiar places to make life easier. He has learned to be very wary of where he is and how to avoid tripping up over other people's chairs or handbags whenever he visits strange pubs or nightclubs. "Once I'd got my bearings I was fine. Toilets were an awful place to find—I couldn't see the sign of the little man on the door or the word 'gents' in small writing. I'd have to wait until I saw a man going in—then I'd know I wouldn't end up in the ladies'!"

One evening in 2012, David and his wife were listening to an interview on BBC Radio News of James Bainbridge, who was talking about a stem cell trial he was getting under way for Stargardt sufferers. They e-mailed Moorfields Eye Hospital that night, and David was soon on his way to London for a battery of tests to make sure he was suitable for recruitment. Bainbridge leads the European arm of an international series of trials sponsored and organized by Advanced Cell Technology, a Massachusetts-based biotech company. They have perfected a technique to harvest human embryonic stem cells from five- or six-day-old embryos, bulk up their numbers, and steer them to differentiate into precursor cells of retinal pigment epithelium (RPE)—the cell layer that lies beneath the photoreceptors, supplies them with nutrients, and carries away their waste products. The science is in its very early stages, and it is not yet possible to inject stem cells into the eye and make them grow into RPE precursors. They have found, to date, that the RPE cell suspensions that work best are almost fully differentiated though not yet fully pigmented. The resident cells in the retina seem able to make the injected cells take the final steps to maturity and cause them to integrate properly with the rest of the RPE and form useful cellular connections. Thus, the phase 1 human trials are testing the safety and effect of injecting 50,000, and later 100,000, cells into the retina of volunteers with age-related macular degeneration and Stargardt disease.

Why are the scientists injecting retinal pigment epithelium cells into patients whose disease involves loss of photoreceptors? Bainbridge explains that the disease process in Stargardt is complicated. A faulty gene in the photoreceptors causes Stargardt disease. It does not stop the photoreceptors from functioning but causes them to dump large amounts of lipofuscin, the debris from worn-out or damaged components of the photoreceptor cells, into the retinal pigment epithelium. Because lipofuscin cannot be degraded, the RPE becomes overloaded with deposits and, because lipofuscin is toxic, the RPE cells start to degenerate. This compromises their metabolic support to the photoreceptors, which, in turn, start to die. It is a vicious cycle of dysfunction and degradation.

Ideally, the best solution would be to use gene therapy, whereby working versions of the faulty gene, ABCA4, could be infiltrated into the nuclei of the photoreceptors using a viral vector. This works by inserting the new gene into the viral genome. The virus then "infects" the photoreceptor and transfers the new gene into it. However the ABCA4 gene is too big for this technology to work, hence the interest in stem cells. But which cells? Replacing photoreceptors without tending to the underlying RPE would be futile, whereas trying to replenish the damaged RPE with precursor cells, without tending to the faulty photoreceptors, can be, at best, a holding operation. Since at the moment they cannot do both, they have opted for the latter. These early trials are to test the "proof of the pudding," and so, instead of injecting RPE precursor cells into the macula, where, because all the photoreceptors have died they could never test for benefit, they locate a patch of peripheral retina where the RPE has died but the photoreceptor population is still largely preserved.

As Hannah Walters, for *The Scientist*, reports, the first trials were done under the jurisdiction of Steven Schwartz at the Jules Stein Eye Institute at UCLA. Human embryonic stem cells were induced to develop into early stage bone and nervous tissue cells, which then differentiated into RPE cells with over 99 percent purity. About 50,000 of these cells were then injected under the retina of two patients: a woman in her seventies with dry macular degeneration and a middle-aged woman with Stargardt. Both were legally blind at the time. The implanted cells certainly survived, and the patients reported some increase in their vision, although the scientists couldn't completely rule out a placebo effect. Some information has been released on the two patients involved, who are both from Southern California. One is a fifty-one-year-old graphics artist from Los Angeles, and the other is a seventy-eight-year-old woman from Laguna Beach. According to a *Washington Post* piece in January 2012, the artist said: "I just woke up one morning and looked through one eye and the other, and the difference was pretty dramatic. I have an armoire across my room and it has a lot of carved detail on it. I looked at it with the eye they operated on and could see all that detail. I just wanted to look at everything. It's sort of like having new eyes." She can now read characters on an eye chart, thread a needle, and see colors. Her sight had started to fade in her twenties due to Stargardt disease. She had lost most of her central vision, and so recognizing familiar faces or watching television became impossible. Now she has regained enough vision to go biking again.

The older woman allowed herself to be identified as Sue Freeman. She suffered from progressive macular degeneration. She had had to stop driving and could not even recognize the faces of family members. "I quit going to the grocery store because I couldn't see any of the labels. I dropped out of organizations I was in. I couldn't take a job because I couldn't read. It definitely changed my life. It shut me down." Within six weeks of the treatment, she started noticing changes. "I started telling my husband, 'Things seem brighter to me. I don't know if it's my imagination but looking at landscapes seems brighter.'" Eventually she got her husband to drop her off at the mall for a trial shop. "He was a wreck, but it was fine." She also realized that she had started reading and cooking again and was able to use her watch to tell the time.

David Lee's transplant operation duly went ahead at the same time that London hosted the Olympic Games. When he came round, his family and James Bainbridge were at the bedside. He was groggy at first and had to lie flat and motionless for nine hours for fear of dislodging the stem cell injection. "It got to one o'clock in the morning, and I could hear all the fireworks going off from the Olympic opening ceremony. It was a good time to come down, London was buzzing—it was a happy place to be, and that made it easier for me." He has no silly illusions that his sight will be miraculously restored, though, in his subjective analysis of the brightness of the colored dots he has to follow in post-operative tests, he thinks there may be some improvement. He's a great sports fan and he can make out the pattern of a football game if he sits very close to the television set. He goes running with two friends in high-visibility jackets and treads in their footsteps to keep out of harm's way. In the back of his mind is always the knowledge that, as Stargardt is genetic, he must have his children's eyes regularly checked. He hopes his volunteer contribution to the science might help the next generation with effective stem cell therapy.

The exciting possibility from Sasai and Eiraku's "eye in a dish" research is that it may soon be possible for scientists like Bainbridge to produce and transplant preformed sheets of RPE cells. The surgery will be more challenging because they will have to make a larger hole in the retina to slot it in, and the substrate on which the cells are grown will have to be permeable to nutrients and metabolites. But it holds promise for more wholesale rejuvenation of RPE, replacing the entire forest rather than a number of trees. And it may only be the start: "It is extraordinary that complex organs like the eye have such powers of self-organization," says Bainbridge. "I think the Japanese work is really remarkable and does demonstrate very vividly the very strong capacity for un-differentiated cells to organize themselves into very complex tissues and even organs. If you could transfer Eiraku's optic cup work into the clinical setting, you would be able to form Müller cells, bipolar cells, glial cells, photoreceptors, and RPE all from one single undifferentiated stem cell stock. So, on the horizon, we could definitely be looking at the wholesale regeneration of retina."

Meanwhile, at the University of Oxford, Robert MacLaren and his team are trying a slightly different approach. They are using so-called pluripotent cells, obtained originally from the skin, to reseed the retina with rods, rather than repair the RPE. This is the research that has been inspired by the results from pioneering patients like Tim Reddish who have received the retinal chip implants. They are secure in the knowledge gained from the implant research that the bipolar cells and the ganglion cells, by which signals are passed to the optic nerve, are still intact even though the photoreceptor layer is hopelessly compromised, and they have obtained exciting results using a mouse model of complete night blindness where all the rods are lost. Injection of many thousands of rod precursors restored vision to the mice. The promise for future human trials is that this therapy will be able to regenerate photoreceptors when retinitis pigmentosa has progressed so far that it has effectively destroyed all the original population.

We've seen how ultra-modern developmental biology has revealed the astonishing extent to which the eye assembles itself because the instructions for making an eye lie within the very developing cells *of* the eye. Has this insight completely demolished creationist arguments against the evolution of the eye? Of course not! Creationists will merely argue that all I have done is hide the hand of God further inside the machinery of eye design. But, in truth, the DIY eye has done enormous service, not only by showing us some important stages in the development of the eye in real time but also convincing science of the powers of self-organization in the embryonic eye—it is the eye that did it itself. It has also supplied ophthalmic scientists with inspiration and raw material so that in the near future they will be able to rebuild diseased eyes. There is now real hope that macular degeneration and retinitis pigmentosa, and a host of other eye diseases, could soon become things of the past.

HOPEFUL MONSTERS

WHY CANCER IS ALMOST IMPOSSIBLE TO CURE

It is ten thirty in the morning in the neurology operating room at Addenbrooke's Hospital, Cambridge, UK, when the fully anesthetized fifty-five-year-old Brian Fearnely is gently and delicately transferred from gurney to operating table, a head-holding clamp already fitted to stabilize his skull in position for the delicate brain surgery that is to follow. Neurosurgeon Colin Watts shaves away a swathe of hair from just above his left ear, over the crown of his head. Using computerized stereotactic technology, he then maps out an area about the size of a computer mouse. A few minutes later the unmistakable acrid smell of singed flesh announces that the Colorado needle, a tungsten-tipped dissecting device, is cutting a precise line through the scalp right down to the bone. The scalp is peeled back, and a section of skull quickly cut through and lifted out. As Watts cuts the dura mater, the fibrous sheet that lies between the skull and the brain, a portion of the brain, about the size of a small hen's egg, wells up proud of the surrounding brain surface. It is a rosy red in color, in contrast to the paler, healthy brain tissue that surrounds it. It is a suspected glioblastoma—one of the commonest forms of brain cancer and the deadliest.

Glioblastomas arise in glial cells, which provide essential physical and nutritional support for neurons in the brain. Watts and his team will typically operate on eighty to a hundred glioblastoma cases every year. It is an extremely aggressive form of cancer with a very poor prognosis. Despite skillful surgery followed by radiotherapy and chemotherapy with temozolomide, the median survival rate is less than five months and less than a quarter of patients survive more than two years. Part of the problem is the difficulty in removing all the cancerous tissue without irreversibly damaging the patient's brain, and the fact that drug delivery to the brain via blood circulation is poor. Relapse—the recurrence of tumor growth—is therefore common. Although Watts has reoperated on some patients, including two reoperations on one patient, he finds that the brain reacts badly to repeated surgical invasion. Patients don't bounce back very well and tend to suffer from cognitive decline, grow moribund, and eventually die.

From a statistical point of view, Peter Fryatt is doing well. He was originally operated on for glioblastoma in October 2011 and proudly shows me the slight crease in his skull that is the only outward sign of the procedure. Despite elaborate care during surgery, operations as intrusive as this can inadvertently cause lasting cognitive damage. "They cut away as much as they could, but a little bit of me disappeared. It's one of those things." As he hunts back to the beginning of his personal journey into cancer, his speech is very slightly slurred. He sometimes has difficulty finding the right word, and his memory for technical detail is not brilliant for a man who, until his illness, was a senior manager in the avionics industry.

His original symptoms were remarkably non-specific. He felt the muzziness you associate with a flu infection, except this went on too long. His GP surprised him by telling him he had diabetes and prescribed some pills. "I said, 'No. No. I think it's more than that and I'd like it scanned.'" The GP prevaricated—scans are costly to the National Health Service—and Peter decided to use his health insurance to have the test done privately because, deep inside, he had an intuition that there was "something else." There was. Two days later the doctors rang him at work to tell him, "You've got a problem—you've got a tumor in your brain." The tumor was quite large—Peter demonstrates at least two joints on his finger.

Two years after the operation, there are many things he used to do but now finds extremely difficult. He doesn't feel as strong or energetic, and his reading is now so slow, he often gives up in frustration and turns on the television. "So I'm watching TV all the bloody time now! And I'm not allowed to drive—that's a pain in the ass too!" And his cancer has relapsed. Scans recently showed two small regrowths, each as big as a fingernail. His cancer is being "re-challenged" with temozolomide to try to knock it out. "Six months ago they saw two bits and they got rid of one, but the other has grown a bit."

His doctors don't know enough to give him a straight answer to "How long have I got?" Reading between their lines, he concludes that it could be anything from tomorrow to ten years. "I'd like to get rid of the tumor, and then I don't have to take these bloody pills anymore. I want to be stronger, walk further, and play golf better than I can at the moment. The doctor tells me how it is—they just don't know the answers."

Back in the operating room, the removal of Brian Fearnely's brain tumor begins in earnest. Several hours before the operation, Brian had been given an injection of 5-aminolevulinic acid—5-ALA, for short. This is preferentially taken up by malignant tumor cells and causes them to fluoresce bright pink when ultraviolet light is shone on them. This allows the surgeon more easily to discriminate between malignant cancer, healthy brain tissue, and dead necrotic tissue. However, in Cambridge, the fluorescence also specifically allows Colin Watts to selectively remove small samples of malignant tissue from different parts of the tumor as he carefully dissects downward into the brain. Within an hour he has removed at least six samples from widely geographically separated parts of the tumor and dispatched them to the cancer genomics laboratory for analysis. Finally, the remains of the tumor are freed and lifted out.

Many of us probably naively believe that cancers are homogeneous lumps of identical "runaway" rapidly dividing cells, but Watts and his fellow cancer researchers in Cambridge know that the reason glioblastoma is such an aggressive tumor, difficult to treat effectively, and with such dismal hopes for survival of affected patients lies in *heterogeneity*. The tumor is not one monolithic block of similar aberrant cells but consists of multiple subpopulations of cells with different genetics, different types of mutations, and different patterns of gene activity. But this heterogeneity has never before been properly investigated because the standard biopsy procedure is based on the taking of one solitary sample from each patient. It is woefully inadequate to reveal any genetic variability that might exist across both space and time, and can never fully sample the whole set of mutations present. This is why they take advantage of the fact that surgical removal of glioblastomas must proceed piecemeal, to take multiple samples from different parts of the tumor and examine each in excruciating detail.

Over the last twenty years, evolutionary biology has invaded cancer research. Scientists who study evolution in cancer see cancers as ecosystems in miniature, composed of a myriad genetically variable entities, or clones, distributed throughout the cancer mass. These clones battle with one another for survival in the same way that animal or plant species compete with one another in the world at large, where climate, nutrients, and other factors act as selection pressures for differential survival, causing evolution to occur. Cancer cells compete for food and oxygen, and have differential resistance to attack by our immune systems and toxic chemotherapy. This selects for those cancer clones that will survive and become the dominant "species." This genetic heterogeneity is synonymous with aggressive malignancy and the more heterogeneous a tumor—the more genetic variability there is between cancer clones—the more difficult it will be to eradicate. The evolution-inspired research on glioblastoma in Cambridge is echoed in research on all cancer types in labs across the world and is leading to answers to the burning questions about cancer: Why does cancer arise in the first place? How do cancers evolve from relatively benign to aggressively malignant? Why do they have a habit of spreading, or metastasizing, to secondary organs or tissues (and why do cells from certain primary tumors have a preference for the organs they metastasize to)? Why is metastasis invariably fatal for the patient? Evolution-think is also already beginning to suggest very novel approaches to cancer treatment.

We are all mutant, claims Mel Greaves, of the Centre for Evolution and Cancer at the Institute of Cancer Research in the UK. If you are middle-aged, just take a good look at your skin. You will almost certainly see that it is peppered with moles and freckles—collectively called nevi. Although most are completely harmless, says Greaves, if you did a genetic analysis on some of them at random, you would certainly find that many contained mutations in a common cancer gene called BRAF, which can lead to unrestrained cell growth. Sample a patch of middle-aged skin, blemished with liver spots, and you will find hundreds of examples of clones of cells containing disabling mutations in an important gene called p53. When working normally, this gene repairs damaged cells or causes them to die off if they are damaged beyond redemption. When disabled, it is powerless to prevent the development of cancer. "If you or I were to go under a body scanner right now, I can bet you they would find all sorts of worrying things," comments Greaves. "I wouldn't do it! So the answer to the question 'Does everybody develop cancer?' is 'Yes!'"

If, perish the thought, I were to fall down dead on the day you read this chapter, a meticulous pathologist dissecting my prostate gland at autopsy would almost certainly discover patches of pre-malignant prostate cancer—called cancer in situ—though this would not be what had killed me. Similar lesions might well be discovered in my thyroid gland, lung, kidney, colon, and pancreas. In Denmark a study of autopsies on women of breast cancer screening age (who had died from non-cancer-related illnesses) showed that 39 percent of them had cancer in situ, which had been completely asymptomatic. Even in children, where the risk of clinical cancer is low at about 1 in 800 between ages one and fifteen, you will find that 1 percent of live births exhibit silent pre-malignant mutations that are the essential founder events for the development of acute lymphoblastic leukemia. This, together with the frequency with which you would also uncover mutations associated with neuroblastomas and kidney cancer, means that one in five newborns may have covert, pre-malignant cancer, says Greaves.

Cancer is, in part, a numbers game. Our epithelial cells and bone marrow, for example, produce 10^{11} cells per day. This rapid rate of cell division means that even with a low mutation rate, it is inevitable that we will eventually accumulate mutations. Our modern lifestyles include a predilection for sunbathing, smoking, eating excessive amounts of red meat, and drinking alcohol. Breast and ovarian tissue is chronically exposed to high levels of female sex hormones in the absence of early and regular pregnancy and prolonged breast-feeding. Cultural trends like these ratchet up the risks imposed by our many evolved design fallibilities, or trade-offs, as in the combination of sunbathing and the gradient toward fair skin at higher latitudes. Our longevity then provides more time for these genetic accidents to happen. "Given this background of mutagenic mayhem fermenting below the surface," Greaves points out, "perhaps the real surprise is that we can live for nine decades with a cancer risk of 'only' one in three." That the cancer rate goes no higher is probably due to the fact that most mutations are either neutral or non-functional; they are "passenger" mutations and not the handful of "driver" mutations for cancer. Even those that do affect either oncogenes or tumor suppressor genes may be the "right" mutation for cancer occurring in the "wrong" tissue or at the "wrong" time to lead to expansion of clones of deviant cells; or will immediately alert other genes that cause the mutated cells to die; or require complementary mutations in other genes acting in concert to cause progression to cancer.

If the frequency of these pre-malignant lesions vastly outweighs the frequency with which malignant cancer occurs, the temptation might be to ignore them completely. The problem is that one in three of us, if we live long enough, *will* be diagnosed with cancer at some point in our lives. Even more chilling, a recent report on cancer estimates that if you were born more recently than 1960, that rate has now risen to one in two. We need to understand why most pre-malignant lesions can sit benignly in organs and tissues for decades, and either regress or do no harm, while others spring to life and develop into life-threatening disease. Unraveling the dynamics of cancer evolution has the potential to overhaul modern oncology. Cancer treatment finds itself caught in a cleft stick because it is constantly at risk of underdiagnosing those cancers that do arise from more benign lesions or overdiagnosing cancer because it identifies pre-malignant lesions and wades in with surgery or chemotherapy for fear that they *will* progress to malignancy.

Greaves's specific area of expertise is leukemia, which is conspicuous for the success with which this suite of cancers can now be treated. Part of the reason is that leukemias tend to be less complex, in terms of the number of mutations required, than most of the solid cancers, and the great success story is chronic myeloid leukemia (CML). This is one of the most straightforward cancers imaginable because it is caused by only one founder mutation. CML, like all forms of leukemia, arises in the bone marrow, where our red and white blood cells develop from stem cells. It affects white blood cells called granulocytes. The most common granulocyte is the neutrophil—a phagocytic white blood cell that travels to the site of infection and gobbles up offending microorganisms. Neutrophils invariably expire on the spot and form the pus we associate with the healing process of stings, cuts, and abrasions.

CML is caused during stem cell division when the ABL gene on the long arm of chromosome 9 accidentally gets transferred to chromosome 22 in an event called a translocation. There it attaches to the BCR gene to make a fusion gene called BCR-ABL. This starts producing a mutated form of the enzyme tyrosine kinase, which normally acts as an "on-off" switch to stop and start cell division. Gene fusion causes the switch to get stuck in the "on" position, and so the cell gets trapped in a situation where it cannot fully differentiate into a mature granulocyte and cannot stop dividing. Eventually, the bone marrow and spleen become choked with these immature cells, and normal production of other types of red and white blood cells becomes prejudiced. CML is treated with a tyrosine-kinase inhibitor (the main one is imatinib, also known as Glivec), which stops this runaway cell division in its tracks. As long as you take it every day—like using toothpaste—you can hold the disease in check for decades, but you can never eliminate it, explains Greaves, because, in the face of medication, the cancerous stem cells just become dormant. If you then take a drug holiday, they will come back and flourish. CML is extremely genetically stable: there is only the one founding mutation and all cells are identical for it, so its very simplicity is the reason why this form of targeted medicine works, although, eventually, further mutations conferring resistance can occur.

In contrast, acute lymphoblastic leukemia (ALL) is more difficult to treat and requires a cocktail of chemotherapeutic drugs, although success rates of over 90 percent are now common depending upon the mutational complexity of the form of the disease. The most common type of ALL affects those stem cells that give rise to B lymphocytes. These white blood cells are one of the most important components of our adaptive immune system because they are almost infinitely variable so that a targeted clone of B cells can be quickly manufactured that is specifically designed to react with the antigens present on the coat of any invading microorganism. As with CML, the initiating event is a fusion gene, this time comprising two genes, ETV6 and RUNX1. This fusion creates a pool of immature B cells that are able to continually renew themselves through cell division. Their accumulation in the bone marrow again puts the production of normal red and white blood cells at risk. This is why affected infants and young children typically present with symptoms of acute tiredness and anemia due to a shortage of red blood cells, bruises and bleeding due to low platelet counts, and decreased resistance to infections due to defective immune systems.

The fusion gene is not inherited, says Greaves, but arises by new mutations, which can occur at any time between week six of the embryo, when it starts making its own blood, and birth. Because bone marrow stem cells are dividing rapidly and mistakes are inevitably occurring at each round of cell division, about 1 percent—one in a hundred—of babies are born carrying the fusion-gene mutation. But the incidence of ALL is far lower, at one case per two thousand, so the vast majority of mutation carriers never experience leukemia. Greaves and his colleagues believe they are now closing in on the mystery as to why the vast majority of fusion-gene carriers never go on to suffer leukemia and why some do. The answer, they think, lies in the brutal weighing of the odds for and against survival by Darwinian evolution, and a crucial mismatch between our exposure to disease pathogens in affluent societies today compared to a century or more ago.

Like the autoimmune diseases we discussed in the chapter "Absent Friends," ALL tracks affluence. It has increased substantially in Western societies since the middle of the last century and continues to rise by about 1 percent per year. Greaves believes that ALL is a "two-hit" disease. The first hit, in utero, is the fusion gene. The second hit is caused by an abnormal immune system response to a florid infection that occurs after we are born and beyond the tender age at which infectious insults to newborn infants used to be more common and are essential to help train and mature their immune systems. This delayed "second hit," in the presence of a deregulated immune system, could produce the stress on proliferating bone marrow cells that triggers a set of critical secondary mutations. Greaves's "delayed infection" hypothesis is therefore an exact parallel to the hygiene hypothesis, which explains our current epidemics of allergic and autoimmune diseases in terms of a lack of early exposure to a wide range of parasitic worms, fungi, and bacteria that would have been endemic in our ancestors' time.

The common type of leukemia in children peaks between ages two and five and is rarely seen past the age of twelve. It is commonly assumed that clones of B-precursor cells containing the fusion gene eventually die out, though no one knows for sure, but Greaves has discovered how these fusion-gene clones can frequently survive until the delayed second hit occurs. The fusion gene activates a molecule called the erythropoietin receptor, which is normally only active in the precursors of red blood cells, where it prevents them dying off and keeps them dividing. It has hijacked a survival mechanism appropriate for another type of cell. A few years later, Greaves theorizes, the affected child succumbs to a delayed infection and the immune system mounts a spirited response. Eventually the body manufactures a cytokine molecule called transforming growth factor-beta (TGF-β) that cools down excessive inflammation because it stops lymphocyte precursors from dividing and so cuts off recruitment of immune cells to fight the infection. But the fusion-gene lymphocytes are deaf to TGF-β. While normal lymphocytes around them fall silent, they keep on merrily dividing and so become dominant in the bone marrow. Thus, delayed infection has rapidly expanded mutant clones at the expense of normal cells, and this expansion is the vital prelude to symptomatic leukemia. Greaves has now discovered precisely how fusion-gene lymphocytes build up cancerous mutations in a process entirely exclusive to lymphoid cells, and which exposes a crucial weakness in evolutionary design.

The key to malignancy, explains Greaves, is the mechanism that has evolved to allow B cells to produce the many antibodies we need to counter the antigens present on the microbes that invade us. The immunoglobulin molecules that constitute our antibodies have hyper-variable regions that can be endlessly shuffled into an infinite number of permutations. About 500 million years ago, our early vertebrate ancestors evolved two enzymes, RAG1 and RAG2, which today target our immunoglobulin antibody genes and shuffle them by mutation to achieve these infinite recombinations. These recombinases are only active in lymphoid cells, and as soon as they have done their job and the cell has stopped dividing and matured into a B lymphocyte, they are switched off. However, in those cases where the fusion gene arises, the cells are constantly dividing, they don't differentiate, and RAG1 and RAG2 are therefore permanently switched on. They very soon run out of immunoglobulin genes to chop and change, and start hunting around for other genes to meddle with. Their precise and temporary mutagenic effects on immunoglobulin molecules are turned into indiscriminate mayhem. So, trapped in an immature, undifferentiated cycle of cell division, the lymphocyte precursor cells gradually build up a suite of further mutations, perhaps a dozen or so, thanks to this off-target action of the recombinases. "Evolution doesn't engineer things to be as specific as you would like to think it should; it just does the best it can," says Greaves, "and, in this case, it has exposed the cells to these off-target effects and caused a major cancer in children. From an evolutionary point of view, this is not a very intelligently designed enzyme. It's a very dangerous enzyme to have."

The hunt for definitive epidemiological proof for the role of infection as the "second hit" has proved elusive because the rate of incidence of childhood leukemia is low in the population as a whole. However, there is evidence from cancer studies in the UK, Scandinavia, and California that attending infant playgroups, which allow for early mixing and exposure to infections, is protective against ALL. In the former East Germany, the Communist government used to run huge infant day care centers to allow mothers to return early to work. However, after reunification, this social engineering was abandoned in favor of home upbringing. Leukemia rates in East Germany had been one-third lower than in West Germany, but after reunification, rates in former East Germany soon caught up.

Twenty years of investigation of "leukemia clusters"—small geographical leukemia hot spots—has done most to convince cancer researchers that the infection hypothesis is correct. One of the most famous clusters occurred in the village of Seascale, near the Sellafield nuclear reprocessing site in Cumbria, UK, where ten times the number of expected cases of childhood leukemia occurred in the years between 1955 and 1973. Fingers automatically pointed toward radiation, but despite finding excess levels of radiation polluting the adjacent Irish Sea, a top-level scientific investigation decided they were not high enough in and around Seascale to cause cancer. It fell to an epidemiologist from Oxford University, Leo Kinlen, to point out that what had changed in Seascale over these years was a large influx of scientists and construction workers to the nearby nuclear site. This substantial population mixing would have late exposed the children in this formerly quiet and remote village to any number of novel infections.

The small town of Fallon, Nevada, is home to one of America's "top gun" air fighter bases. Between 1999 and 2003, thirteen cases of childhood leukemia were reported, where, statistically, you would have expected less than one. The locals grimly suspected that spills and dumping of JP-8 fuel, a carcinogenic mix of kerosene and benzene, were to blame, citing the fact that the fighter aces slurped their way through 34 million gallons of fuel in 2000 alone. However, the official study report found it could not successfully pin down any environmental contaminant. What had changed was population number. The local resident population of Fallon is 7,500, but this had swelled to a fluctuating population of 20,000 in the 1990s and a whopping 55,000 in 2000, thanks to influx of military personnel, builders, and logistics and support workers.

Greaves is currently investigating a leukemia cluster in a primary school in Milan. "There are seven cases, which doesn't sound that much, but four cases occurred in the same school in just one month shortly followed by three more—that's out of sight. You'd expect one case in fifteen years, at most, in a school that size." The fact that the children varied in age between three and eleven yet all came down with leukemia at the same time suggested a shared environmental trigger. Greaves's team tracked back a few months before the outbreak to find that there had been an epidemic of swine flu at the school. Whereas the swine flu infected one in three in the general population, all seven leukemia cases had become infected. "So the stats were not too great because of the small sample size, but it is highly compatible with swine flu being the 'second hit.'" The only other supporting evidence, says Greaves, comes again from epidemiology from Oxford that tracked the incidence of ALL over thirty years throughout the UK. There are two peaks, and they both come within six months of a seasonal flu epidemic.

Our susceptibility to cancer goes back over a billion years to the origin of the first multicellular animals. Before that, life was unicellular and all cells were free to selfishly reproduce at will. But with the advent of multicellularity, cells had to learn how to work together. They could no longer divide ad nauseam, and cell division became tightly constrained to stem cells and the immediate progenitor cells that descend from them and have a more limited repertoire for both replication and differentiation. These progenitor cells have a limited life span, which means that a cancer-inducing mutation occurring in one of these cells is likely, therefore, to meet an evolutionary dead end because its prospects of expanding into a clone of cancer cells is severely limited. Once a cell has fully differentiated into a muscle, liver, or skin cell, for instance, it loses its immortality altogether. Immortality is therefore restricted to small pools of stem cells that are necessary for embryonic development, a constant supply of red blood cells and immune system cells, and the regeneration of tissues and organs caused by age and wear-and-tear. This suggests that, in order for a cancer to originate, mutations must either target stem cells, as in the leukemias, or specific mutations should occur in differentiated or semi-differentiated cells that return them to an immature state in which they can reenter the cell cycle and resume cell division.

This new era of cooperation and conformity required the evolution of genes and chemical signaling pathways within and between cells to police and enforce the new rules. Additional mechanisms of DNA repair evolved to intercept cancer-causing mutations. If DNA damage exceeded a certain point, these new genes would initiate the death of the cell, and they are therefore called tumor suppressor genes. Further suppressor genes evolved that stopped cells dividing by blocking mitosis, the process by which cells divide and replicate. These are called cell cycle checkpoint genes. Vertebrates also evolved sophisticated adaptive immune systems. In the same way that these can produce clones of tailor-made lymphocytes to counter the antigens present on the coats of bacteria and viruses, they can also target cells that are turning into cancer.

As Matias Casás-Selves and James DeGregori, from the University of Colorado, explain, the very evolution of animals—their tissues, organs, and systems—has been constrained by the need to avoid cancer, which is why they have evolved such potent tumor suppressor mechanisms. We have had to limit the growth of rogue cells that had refused to conform to the new rules of multicellularity and have placed a number of robust barriers across the road to cancer development, barriers that cancer cells have to overcome if they, in turn, are to survive and multiply. In 2000 American cancer researchers Douglas Hanahan and Robert Weinberg condensed these properties down to a list of what they call the six hallmarks of cancer, but you can equally see them as the six hurdles that cells have to jump if they are to become cancerous.

First, they explain, cancer cells had to evolve self-sufficiency in growth signals. Normally, cells receive signals from outside that are transmitted across their cell membranes by receptors. These growth factors wake the cell up from its quiescent state and stimulate it to begin dividing. Cancer cells can produce their own growth factors that mimic control from outside. Two common examples are platelet-derived growth factor (PDGF) and transforming growth factor-alpha (TGF-α). They can also increase the activity of the growth factor receptors on their membranes by vastly increasing the number of gene copies for any particular receptor. This has the effect of making the cancer cell overreactive to ambient levels of growth factor that might not otherwise trigger cell division. Two classic examples are epidermal growth factor receptor (EGFR), very common in brain cancers, and the human epidermal growth factor receptor 2 (HER2) in breast cancer. They can also produce mutated forms of RAS proteins that jam in the "on" position the signals that stimulate the cell to divide. Second, would-be cancer cells must turn a deaf ear to antigrowth signals. A classic example is the development of insensitivity to transforming growth factor-beta (TGF-β) that we see in acute childhood leukemia.

Third, cancer cells must be difficult to kill. Normally, when mutations occur, or damage to chromosomes is detected, repair mechanisms in the cell leap into action. If the damage is too great, the cell undergoes a process called programmed cell death, or apoptosis. This is where the tumor suppressor gene p53 comes in because it is an important trigger for apoptosis through sensing DNA damage. Cancer cells must disable genes like p53, or they will be dead inside half an hour as their cell membranes are disrupted, their internal structure destroyed, the nuclei fragmented, and the chromosomes degraded beyond use. Their shriveled corpses, explain Hanahan and Weinberg, are engulfed by nearby scavenging cells and disappear within twenty-four hours.

If cancer cells are to give rise to the colony we call a tumor, which can contain upward of 1 trillion cells, they must become immortal through a limitless ability to divide and double their number. Normal differentiated cells in the body, like heart cells, cannot divide any further, but many cell types, like skin fibroblasts, retain a limited repertoire of cell division and can be encouraged in cell culture to undergo a number of doublings before senescence sets in and the cell enters a crisis state of massive chromosomal disarray from which it cannot recover. Cancer cells must tap into a mechanism to avoid this if they are to achieve true immortality. In normal cells there are repetitive DNA sequences at the tips of chromosomes called telomeres that protect the DNA inside them from damage. These telomeres are gradually whittled away with each successive round of cell division, which eventually leads to fatal degradation of chromosomes, and the cell dies. But cancer cells increase the activity of an enzyme, telomerase, which is strongly suppressed in normal cells. This rebuilds telomeres as fast as they are shortened, constantly maintaining them and endowing the cell with unlimited powers of replication.

No cell, cancerous or otherwise, can survive without access to oxygen and nutrients. A cancer cell is approximately 20 microns (a micron is a millionth of a meter) in diameter. If it lies more than 150 microns away from a blood capillary, it will die. This causes a bottleneck in the proliferation of cancer cells in a tumor because the production of new blood vessels, called angiogenesis, is tightly regulated. A cancer clone, therefore, must acquire mutations that allow it to stimulate the production of new blood vessels, and cancers commonly do this by causing the increased production of vascular endothelial growth factor (VEGF) either by activating the RAS oncogene or disabling the tumor suppressor gene p53. Consequently, cancer researchers frequently see this loss of p53 in midstage cancers, before they develop into full-blown malignant tumors.

Finally, the key to immortality of cancer cells (at least as long as the patient survives) depends on them developing the ability to detach themselves from the initial tumor mass and travel to other sites in the body, where they can begin development into secondary tumors. This process is called metastasis and is responsible for at least 90 percent of all cancer deaths.

"How are normal cells," ask Casás-Selves and DeGregori, "which are part of a strict tissue organization with cellular social cues, transformed into cellular sociopaths who disregard tissue order and ignore cellular dialogue?" The granddaddy of cancer evolution is Peter Nowell, now emeritus professor at the University of Pennsylvania. As long ago as 1976, he detailed the route from a normal, benign cell to a malignant cancer. It was Nowell who was among the first researchers to point out that, over time, cancers tend to increase their ability to proliferate by escaping further and further from the control mechanisms by which normal cells abide. And as they proliferate and become more malignant, they lose differentiation. They jettison the organelles and metabolic functions that allowed them to function as a specialized cell and revert to a simpler type of cell that concentrates all its energy on proliferation and invasive growth. They do this, said Nowell, by accumulating a set of mutations that allow them to turn a deaf ear to the body's control systems and allow them to begin unlimited cell division. This results in a neoplasm, or mass of cells, that have mutated to gain some advantage over surrounding cells. Cells in this initial clone can independently accumulate further mutations and split into new clones, within the tumor mass, with different properties, levels of malignancy, and susceptibilities to treatment.

According to Mel Greaves and Carlo Maley (a prominent evolution in cancer researcher from UC San Francisco), thirty years of research since Nowell's early descriptions of the cancer process have thoroughly borne his ideas out. "A large body of data from tissue section, small biopsy, and single cell analysis supports Nowell," they say, "in that it shows that the evolutionary trajectories that arise are complex and branching providing a striking parallel with Darwin's iconic evolutionary speciation tree. Divergent cancer clones in this context parallel allopatric speciation in separated natural habitats—as in the Galápagos finches."

Darwin likened the evolution of life on Earth not as a linear process, but as an endlessly branching process with each species alive today represented by the terminal twigs on a tree's multiple branches. The evolution of cancer clones, within the mass of one single tumor, is Darwinian evolution in miniature. And as you can trace backward from a twig to the base of a trunk, it is possible to trace the common origin of cancer clones even if they have accumulated sufficient changes to be wildly different from one another. The Galápagos Islands provided a wonderful living example of how speciation can occur when breeding individuals from a founder population, as in the famous Galápagos finches, became geographically separated (allopatric) on the many islands in the archipelago. The microenvironment inside a tumor, and in its immediate surroundings, provides exactly the same level of environmental heterogeneity thanks to widely different levels of blood supply, oxygen, nutrients, competition between clones, and predation by the immune system.

Different cancer types take very different routes to malignancy. Nowhere is this better demonstrated than inside the human gut, where there are at least four major types of colorectal cancer. Joe Weigand has been branded an ultramutator because he has suffered from a relatively rare form of colorectal cancer that becomes mutation-mad. The condition is passed from parents to their children, in contrast to the majority of cancers that are sporadic and depend on de novo mutations occurring in any one generation. Joe, and his doctors, had plenty of warning that he could be facing trouble. His paternal grandmother had died from colon cancer in her mid-forties, and his father was diagnosed with suspected colon cancer at the same age. By this time, endoscopy could be offered to those at risk, and examination of his father's colon revealed hundreds of pre-cancerous polyps. Joe and his sister were still toddlers when their father heard the bad news, and he was unwilling to put his young family at risk by waiting to see if any of the pre-cancerous polyps underwent worrying changes. He feared they might evade surveillance and develop into full-blown cancer. He therefore agreed to have his entire bowel removed and has been living with a colostomy bag ever since.

Not surprisingly, Joe was checked out regularly, but a budding career in the hurly-burly financial sector led him to cancel his regular colonoscopies for four years. Alarming weight loss drove him to his GP. "I'd lost about 30 to 40 percent body weight—I looked like a ghost. I was completely anemic and had no energy." His GP ignored the family history of cancer that was staring him in the face and prescribed iron tablets for anemia. "I was in London, living with my brother, and my dad came up to visit us one day, took one look at me, and said, 'This is ridiculous, bugger the NHS, I'm going to pay for you to go and have private exploratory stuff!' And that's when the colonoscopy found them." During his surgery four weeks later, they discovered twenty or thirty small polyps and one huge tumor the size of a mango. They removed it together with the vast majority of his colon. "I've literally got thirty or forty centimeters left, but it means I can still go to the toilet normally." Now, eight years post-operation, he leads a normal active life but still has to have periodic colonoscopies at which small polyps are regularly found. "They found four polyps last week. They tend to find some more every time they go in. They just zap them there and then, and take them out for histology. When they're small they tend to look quite normal, it's when they get bigger that they go a bit haywire and accumulate a lot of mutations."

The scientist who has investigated this type of cancer, Ian Tomlinson from Oxford University, has christened it with a mouthful of a name—polymerase proofreading-associated polyposis. When DNA copies itself, to form the genetic content of daughter cells, it occasionally makes mistakes and inserts the wrong DNA base into the genetic code. There are two polymerase enzymes that detect these mistakes and repair them. When the two genes that code for those enzymes are mutated, at least half the mistakes go unnoticed so that these tumors characteristically accumulate over a million mutations compared to a range of between ten and several thousand for most cancers. The outcome for patients is very variable, however, because sheer number of mutations does not necessarily equate with malignancy. It is far from clear what handful of genes, among the million that become mutated, actually drive the cancer, but these cancers are not particularly aggressive and the mutational load, rather than leading to increased malignancy, may actually disable so many functions in the cancer cells that many of them die off.

Joe's colon cancer is in stark contrast to colorectal cancers that occur mainly in the distal colon, toward the rectum, and which tend to be far more malignant. These colon cancers do not have excessive rates of mutations within individual genes because their DNA repair mechanisms remain intact. Instead, they typically exhibit very extraordinary levels of chromosomal instability, a feature they share with the vast majority of cancers, whereby whole chromosomes, or major arms of chromosomes, easily capable of containing hundreds of genes, suffer gross structural abnormality. Recent cancer research has established that this chromosomal instability underpins malignancy and is much more important than the role of simple point mutations, per se, in the genetic code.

The total complement of chromosomes in the nucleus of the cell is known as the karyotype. With a number of specialized exceptions, all normal body cells are diploid, containing twenty-three pairs of chromosomes, one of each pair descending from the mother, the other from the father. But the vast majority of malignant cancer cells have been found to exhibit wild departures from this normal state of ploidy, all caused by faults in mitosis—the method by which chromosomes divide and segregate to form two daughter cells.

Mitosis involves the duplication of each chromosome to form two sisters. As the cell envelope elongates and the cytoplasm begins to divide to produce two identical cells, the sisters assemble at the center on a structure called the centromere. A series of protein microtubules, called the mitotic spindle, then forms, which attaches to each sister chromosome and converges at two opposite poles. The sisters are then drawn apart along the spindle until they congregate at the opposite poles and condense to become the nuclei of the new daughter cells. Anything that interferes with the absolute fidelity with which this process proceeds is likely to result in whole chromosomes or bits of chromosomes failing to arrive at their destination. Abnormal mitosis can result in hypodiploidy, where there are considerably fewer than 46 chromosomes, or tetraploidy, a doubling of chromosome number. Chromosomes that have either lost or gained copies are collectively termed aneuploid.

The pathologist David von Hansemann first reported this aberrant process in samples of cancerous tissue in 1890. He was followed by Theodor Boveri, a zoologist researching in the early twentieth century, who first pointed out that mis-segregation of chromosomes caused by abnormal mitosis led to aneuploidy and might be a cause of tumor development by occasionally giving rise to a malignant cell with the ability for "*schrankenloser Vermehrung*"—unlimited growth. As Zuzana Storchová and Christian Kuffer point out, this "old" theory of chromosomal instability took a backseat, during the explosion of the genomics era, to the idea that mutations were the most important events that caused cancer. But it has shot back to prominence in the last few years as researchers begin to realize that unstable chromosomes are not simply the background genomic havoc that mutations are capable of creating but that the opposite is true—genetic instability is part and parcel of the process of generating cancer-causing mutations, cancer clone diversity, malignancy, and metastasis. In fact, in most cancers, chromosomal instability and mutation go hand in hand. Mutations allow chromosomal instability to occur in the first place, and the instability further increases the mutation rate.

How could tetraploidy—chromosome doubling—lead to malignancy? It could simply be that it allows the cell to survive while it is undergoing a spate of mutations that might otherwise prove fatal. Working genes might be preserved among some chromosome copies while their sisters are being dismantled and functionally changed by mutation. But it is also a major route to the irregular aneuploidy that typifies most cancer. A cancer cell might first become tetraploid and then whittle away at its genome over time so that whole chunks, arms, and, sometimes, complete chromosomes are simply lost.

Aneuploidy can result in both losses and gains of genes. If a whole chromosome or a part of a chromosome goes missing, the genes on it will be lost. Since all genes exist in pairs, called alleles, this would leave only one allele of any particular gene on the sister chromosome that remained unscathed.

The remaining allele is therefore completely exposed to further mutational change that might lead to the total loss of that gene. When this happens to the tumor suppressor gene p53, for example, this allows the mutant cell to ignore signals instructing it to die.

Aneuploidy can also give rise to translocations, when parts of chromosomes mistakenly attach themselves to others and either give rise to the types of fusion genes encountered in the leukemias, or lead to massive increases in the number of copies of genes—a process called amplification. Gains and losses of alleles are called copy number changes. The results can be drastic. The average cancer of the colon, breast, pancreas, or prostate can lose 25 percent of its alleles, and it is not unusual for a tumor to lose over half its alleles. One study showed that gains or losses of multiple chromosomes occurred between ten and a hundred more times in aneuploid colorectal cancers than in normal cells or diploid cancers of the same type.

There is a long list of genes that, when mutated, seem to initiate chromosomal instability in cancers. They are genes that promote cell proliferation, disorganize mitosis, or prevent cancer cell euthanasia. For instance, BRCA1 and BRCA2, the two genes that predispose to breast cancer, are among those genes that repair DNA and regulate cell division; BUB1 and MAD2 organize the assembly of chromosomes on the mitotic spindle; APC gene mutations are often seen early on in the development of colorectal tumors, and this gene also seems also to be involved in formation of the mitotic spindle and the division of cytoplasm to form daughter cells; and, as we have seen, p53 mutations could silence a gene that would normally either attempt to repair the DNA damage that occurs in aneuploidy or kill the cell off. In truth, say Christoph Lengauer and Bert Vogelstein, two veterans of evolution in cancer research, there are so many genes that can, when mutated, provide the affected cell with the instability it requires to develop further multiple genetic alterations that lead to malignancy. Chromosomal instability is the engine of tumor progression and tumor heterogeneity, guaranteeing that no two tumors are exactly alike and that no single tumor is composed of genetically identical cells. This is the nightmare for oncologists: it is why they are always chasing shadows, and it is the main stumbling block to any truly successful cure for cancer.

This is exactly the pattern that Colin Watts and his colleagues have found in Cambridge with glioblastoma, and it explains why Peter Fryatt's cancer is relapsing and why his prognosis is so uncertain. Deep molecular analysis of the genomes of cells in each sample fragment of one glioblastoma allowed them to identify a founding clone that had accumulated a number of mutations and chromosome instabilities. This had then split into a complex branching tree of clones that increasingly diverged from one another and accumulated different malignant traits. An early event in tumor evolution was a chromosomal instability that resulted in the formation of a highly aberrant circular chromosome called a double-minute, which could replicate under its own steam and contained hundreds of copies, instead of the normal two, of the EGFR gene, which influences cell proliferation and migration. This clone had also gained copies of the MET gene, which gives rise to invasive growth and a poor prognosis for the patient, and had lost copies of the tumor suppressor genes CDKN2A and PTEN. This clone then split into two subclones, which gained parts of one chromosome and lost parts of others, accumulated further mutations in tumor suppressors, and finally diverged into five substantially different tumor clones.

Chromosomal instability in cancer can occasionally become so abrupt and outrageous in its proportions that the researchers who study it are led to suggest that cancer evolution does not conform to classical evolutionary theory. In 2011 Philip Stephens and a group of fellow researchers mainly from Cambridge, UK, reported their discovery of a cataclysmic event that had caused hundreds of chromosomal abnormalities to arise in one fell crisis in a white blood cell from a sixty-two-year-old woman with chronic lymphocytic leukemia. They coined a new name for it—chromothripsis—meaning "chromosomes shattering into pieces." The event had occurred before the woman had been diagnosed with cancer and had resulted in a clone of cells that were resistant to alemtuzumab, a monoclonal antibody commonly used to treat this kind of leukemia. Her condition had deteriorated rapidly as a result. They totted up forty-two genomic rearrangements on the long arm of chromosome 4 alone, plus further rearrangements on chromosomes 1, 12, and 15. These had caused huge numbers of copy number differences in genes, characteristically with one copy being lost. These losses were not down to simple deletion, said the researchers, but to a vast number of chromosomal breakpoints coinciding with the position of genes. When they examined each breakpoint in detail, they frequently found that the two regions of chromosome that had rejoined at the breakpoint had not originally been next to each other. The chromosome had literally shattered, they surmise, leaving hundreds of shards of DNA circulating unfettered in the nucleus. This caused the DNA repair machinery to leap into action, like a whirling dervish, pasting bits back together willy-nilly "in a helter-skelter tumult of activity." "The resultant hodge-podge," they say, "bears little resemblance to the original chromosome structure and the scale of genomic disruption has wholesale and potentially oncogenic effects."

Nor was this an isolated occurrence. They noted a similar chromothripsis in lung cancer cells. Chromosome 8 had shattered into hundreds of pieces, which were stitched back together again into a patchwork chromosome with the exception of fifteen DNA fragments that had joined together to form a highly abnormal circular double-minute chromosome (similar to the one identified in the glioblastoma research) containing up to two hundred copies of the MYC oncogene. This massive amplification had conferred a huge selective advantage on this cancer cell line and had made it more malignant. They have seen the same pattern of chromothripsis in many cancer types, including lung cancer, glioma, bone marrow, esophagus, colon, and kidney. It is particularly common, and profound, in bone. The key question here, for evolutionary biologists, is whether to view chromothripsis as an entirely drastic random event involving the explosion and remodeling of a cancer genome, that, by accident, in one in a billion cases, confers a competitive advantage upon a cancer cell, or whether it is actually happening in a non-random way as a dire strategy—or mechanism—to confer a selective advantage on that particular cancer clone in the face of extremely strong selection pressure.

For instance, a group of British researchers have been taking a hard look at a type of childhood acute lymphoblastic leukemia that is caused by another dramatic chromosome fusion event—that of chromosome 21 with chromosome 15 in a so-called Robertsonian translocation. Individuals born with this rare chromosomal event have a 2,700-fold increased risk of developing this type of leukemia. Following the translocation, the unstable fused chromosome shatters in a dramatic chromothripsis event and, again, is pasted back together in a seemingly random fashion by the cell's DNA repair enzymes. The researchers believe that it is the structural abnormality of the fused chromosome that actually makes chromothripsis more likely.

You might expect that cancer clones carrying such an unstable patchwork chromosome would be at a disadvantage in competition with other clones in the tumor—in which case chromothripsis would be an evolutionary dead end. But although this cataclysmic event *appears* random and chaotic, it specifically spares and amplifies certain regions of the chromosome, increasing gene copies and the activity of several genes that are known to be involved in malignant blood cancers, including RUNX1, DRYK1A, and ETS2.

In the final stage of chromosome transformation, the whole fused chromosome is invariably duplicated and occasionally turned into an aberrant ring chromosome containing multiple copies of these malignancy-causing genes. The researchers then compared the pattern of gene-copy number changes in the fused chromosome with samples of chromosome 21 from a wide range of cancers and found that it closely mirrored them. Remarkably, they note, chromothripsis might be remodeling chromosome 21 in a non-random way so that the genes that cause it to develop greater malignancy and drive it toward leukemia can thrive.

That all this occurs in one cataclysm, rather than via the gradual accumulation of mutations, suggests to all these researchers that, in the peculiar world of cancer cells, evolution often proceeds by what is called punctuated equilibrium. This theory—developed by Niles Eldredge and Stephen Jay Gould in the context of the evolution of species of multicellular animals over periods of geological time—maintains that evolution proceeds via long periods of evolutionary stasis that are occasionally interrupted, punctured, by sharp bursts of mutational change, before subsiding once more into stasis. Their ideas challenged the idea that is favored by the majority of evolutionary biologists—the gradualists—that evolution plods along through a stepwise accumulation of favorable mutations.

However, the cataclysmic genetic change that occurs in chromothripsis brings to mind an even more outrageous theory as to how evolution—and speciation—can occur. This was championed by the evolutionary heretic Richard Goldschmidt and branded the "hopeful monsters" theory. Goldschmidt was a formidable geneticist and a Jewish refugee who escaped the Nazis to take up a position at the University of California, Berkeley. He crossed the line, for the vast majority of evolutionary biologists, with the publication in 1940 of his book *The Material Basis of Evolution*, which maintains that gradual stepwise accumulation of mutations are not sufficient to explain the evolution of one species from another; macro-mutations—drastic mutational change—are necessary. Eventually the gradualists won the day, arguing that such wholesale genomic change would invariably prove disastrous for complex multicellular organisms. Goldschmidt's ideas were consigned to oblivion. But it has become obvious in the intervening years that in much simpler organisms like many bacterial species, sudden massive genomic change can occasionally pay off, despite the loss of many unviable individuals, if the selection pressure on organisms is sufficiently harsh. Some cancer researchers are beginning to evoke Goldschmidt's name because of the wholesale, seemingly catastrophic mayhem they are discovering in cancer cell nuclei. In the specific context of cancer evolution, Goldschmidt is being rehabilitated.

Charles Swanton, from Cancer Research UK's London Research Institute, has shown exactly how tetraploidy—genome doubling—causes chromosomal instability and accelerates tumor evolution in colon cancer. In a group of 150 patients with colorectal cancer, genome doubling increased the chance of relapse by five times at two years after treatment. By following several tetraploid clones of colorectal cancer in cell culture, his team was able to show marked differences between the genomes of tetraploid clones and diploid cancer cells, including the loss of large regions of chromosome 4, which predicted these very poor patient outcomes. Swanton is convinced that this genome-doubling event in tumors, like chromothripsis, represents an evolutionary "hopeful monster" leap in cancers that was originally called saltation by many early eighteenth- and nineteenth-century biologists, whereby organisms are proposed to evolve abruptly and drastically by a major macro-mutational step. Prior to Darwin, most evolutionary scientists were saltationists, including Etienne Geoffroy Saint-Hilaire and Richard Owen, who both proposed that monstrosities could become the founding fathers of new species through instantaneous transition from one form to the next. Richard Goldschmidt—branded a heretic for his theory of "hopeful monsters"—was a twentieth-century saltationist. Evidence is now piling up that tetraploidy, aneuploidy, genetic instability, and rapid and substantial changes in the copy number of genes all have the potential to produce saltational change and drive cancer evolution.

Cancer cells are indubitably monsters. Their bloated, abnormal nuclei are obvious when you look at them through a microscope, and numerous advanced molecular biology techniques reveal the profound abnormality of their genomes. But they are hopeful monsters in the sense that the odds against their survival are absolutely enormous. As Greaves and Maley point out, the doubling time of cancer cells is of the order of one to two days, but the doubling time of a tumor is between sixty to two hundred days. This indicates that the majority of cancer cells die before they can divide. But, occasionally, one catastrophic event survives by conferring some crucial survival advantage. As Greaves puts it, "In conditions of severe environmental stress, one way cancer cells try to escape or adapt is just to scramble everything—go for instability. This is Goldshmidt's 'hopeful monster' idea. The game plan is 'OK I'm going to throw everything up—one will survive whereas 99.9% of cells will die.'" If that cell manages to give rise to a clone of similar deranged cells, a dangerous cancer will have evolved.

Trevor Graham, presently working at the Barts Cancer Institute of St. Bartholomew's Hospital in London, is researching a form of colorectal cancer that arises in just such a harsh environment, the large intestine of sufferers from inflammatory bowel disorders. Here the selective force for cancer comes from the damage that occurs to the mucosa—the layer of the gut wall that comes into contact with gut contents. As any sufferer from ulcerative colitis will tell you, from rueful inspection of their frequent, painful, bloody bowel movements, their large intestine becomes incredibly damaged, with a great deal of killing and removing of mucosal cells. "So there's huge selection," notes Graham, "for being hardy in the horrible place that is the inflammatory bowel." But ulcerative colitis, like other inflammatory bowel disorders, is episodic. Following acute episodes where the mucosa is flayed alive, there is a healing period where it is rebuilt. Any cells that have managed to survive the inflammatory holocaust will be those that proliferate during this repopulation, and if they are able to divide faster than normal cells, pre-cancerous clones, already containing a high frequency of cancer-predisposing mutations, can take over vast geographical areas of colon. Such clones invariably develop chromosomal instability and can often be over a meter long. Graham has recorded one clone that spread all the way from the top of the colon to the rectum.

However, even if a pre-cancerous clone takes over a large part of the intestine, this does not automatically lead to malignancy. Gastroenterologists monitoring their patients with endoscopy will notice large patches of dysplasia—pre-cancerous lesions—which they term low grade or high grade according to the morphological appearance of the cells down a microscope. As Trevor Graham explains, high-grade dysplasias are more likely to progress to cancer, and so about 50 percent of patients with high-grade dysplasia will get cancer in the short term. But this means 50 percent will not. Worse still, identification of dysplasia grade is a notoriously subjective process and estimated likelihood that a dysplasia will progress to cancer varies from 2 to 60 percent depending on whom you ask. "So everyone with high-grade dysplasia is offered a colectomy," says Graham, "and then you don't have a colon anymore and you have a stoma and a colostomy bag."

At the moment it is beyond the reach of oncology to discriminate between those patients who are at high risk for developing cancer and those at low risk. As Graham puts it: "The biggest contribution to quality of life in all this is sparing the low-risk patients from massive over-intervention. That means sparing more of them from colectomy, not treating them, and being confident in that decision, which we can't do at the moment because no one wants to take the risk."

The holy grail for cancer researchers will be to discover a dependable way of discriminating between those pre-cancerous clones that will likely progress to malignancy and those that will not, thereby saving a lot of patients from unnecessary surgery. But the bewildering level of clonal heterogeneity in cancers, involving chromosome abnormality and instability, and the different complex permutations of mutated genes that are associated with malignancy all defy such prediction. The cancer researcher who has done most to inject an element of certainty into this uncertain world is Brian Reid, of the Fred Hutchinson Cancer Research Center in Seattle. Together with Carlo Maley and many other colleagues, he has been studying a pre-cancerous lesion called Barrett's esophagus and has built up one of the biggest cohorts of Barrett's patients in the United States. He has been monitoring the progression of their disease in extremely fine detail over decades. If the problem with overdiagnosis is acute with regards to bowel cancer, it is even more so with Barrett's, where the rate of progression of even high-grade dysplasia to malignancy is only 15 percent over ten years. As Reid explains: "The current recommendations for Barrett's esophagus are once-in-a-lifetime screening at age fifty. That detects those dysplasias that are going to remain stable for life. Patients will die with it, not because of it. It misses those that are going on to cancer even though we know the cancer arises in Barrett's. So we have overdiagnosis of these very benign cases that will not transition to cancer, and that's about 95 percent, and we have underdiagnosis of cancers that are going to kill people, and that is also 95 percent. We're doing very badly in that regard."

It is an example, says Reid, of length-time bias in cancer screening by which all screening tests selectively detect slow-growing or non-progressive conditions and selectively miss rapidly progressing conditions. Esophageal cancer rates, notes Reid, have been on the rise in the West for the last four decades. It is the fastest-growing cancer in the United States, and it is now beginning to be seen in Asia, where once it was very rare. Obesity triples your chances of getting esophageal cancer. So does a combination of heavy smoking and drinking spirits (as the author and polemicist Christopher Hitchens discovered to his cost) and a diet low in fruit and vegetables. The cancer is highly lethal with a survival rate of only 15 percent. Barrett's esophagus is a change in the appearance of epithelial cells caused by acid reflux from the stomach. The chronic inflammation caused by stomach acid can lead to Barrett's and initiate the progression of Barrett's to adenocarcinoma. The preventative route has always been to screen individuals with long-term reflux and, if they have Barrett's, to treat it aggressively before cancer develops. At its worst this means removing the whole esophagus, a difficult and highly dangerous procedure that, until recently, only about 20 percent of patients survived. Reid saw the necessity for long-term detailed monitoring of Barrett's patients in the hope of discovering a way of predicting the relatively small percentage of Barrett's patients who were likely to get cancer.

Lewis Quierolo was first diagnosed with Barrett's in 1989. He had been suffering from acid reflux for the previous ten years. While at college he was a competitive weightlifter, and this had caused a hiatus hernia that made his reflux even more severe. "I would have a bottle of Maalox by the side of my bed and get up three or four times in the night to take a full-mouth swallow to put the fire out so that I could get back to sleep." His doctors put his acid reflux down to a student lifestyle and diet. "They told me to elevate my head a few inches when going to sleep—it might help. But I slept on a waterbed and so raising one end of that accomplished absolutely nothing!"

In 1989 he finally did see a gastroenterologist who identified Barrett's and also volunteered the information that there was a study of Barrett's going on at the nearby University of Washington, led by Brian Reid, who was holding a "closed-doors" meeting of clinicians the following day. Lewis decided to crash it. "I introduced myself as Dr. Lewis Quierolo and just walked in. I tracked Brian down over pre-conference refreshments, introduced myself, and owned up that I wasn't actually a physician. Brian very generously pulled me aside and said, 'I really wouldn't recommend you come in and listen to my talk because you'll be frightened to death!' He explained that he would be talking about the tiny percentage that go to the extreme outcome and showing graphic slides." Lewis stayed and volunteered for Reid's study. Reid performed an endoscopy on him and signed him up straightaway. He has remained in the study, with no progression to cancer, regularly monitored for twenty-five years.

Author Bob Tell, from Michigan, had a far more frightening experience with the onset of Barrett's and came within a whisker of losing his esophagus. He was diagnosed with Barrett's esophagus, high-grade dysplasia (cancer in situ), at age sixty. He had stoically suffered from acid reflux for thirty-five years but "toughed it out" by taking common over-the-counter antacids. "I ate poorly at the time because I happen to like eating junk—I've paid the price for all those hamburgers, pizzas, and curries!"

His diagnosis was almost accidental. He was having a routine colonoscopy to see if there were any growths in his bowel when the gastroenterologist said, "You've complained of heartburn. As long as we are going to have to put you under, we might as well check the other end!" They sent the esophageal biopsy slides over to an expert pathologist at the University of Michigan who gravely responded that the cell pathology was so worrying they would have to act fast. Bob immediately came under enormous pressure from all sides to have an esophagectomy. His gastroenterologist predicted a rapid, agonizing death within four years; his cousin, who was chair of oncology at Johns Hopkins University, said, "You poor guy—that's too bad—but you can't take the chance"; and a pathologist friend told him that if this had happened to him, he would have leapt onto the operating table.

Bob was between the proverbial rock and a hard place because he also knew that, at the time, the statistics for esophagectomy were terrible and the quality of life afterward very poor. "When God designed the human body, he didn't intend surgeons to get at the esophagus—the mortality numbers were atrocious." A female friend had had the operation done; they had removed the esophagus, pulled her stomach up into her chest cavity, and attached it to what was left. She had no swallowing reflex anymore, had to eat in very small amounts, and was always rushing to the bathroom to regurgitate. He thought, "There simply has to be an alternative to all this" and hit the Internet.

Before long he came across Brian Reid's study in Seattle and, in a long shot, rang them up. "Brian's assistant Christine was very persuasive. She said 'Just come on out—come on out next week! It might just be inflammation, or low grade. Let's at least take a look at you!' I thought, 'What harm is there?'" Several weeks later, quaking with terror, he was sat in front of Reid. "Brian was wonderful. He quickly eased my fears. He outlined, within an hour or so, the entire development of cancer out of Barrett's dysplasia, how slowly it grows, and why the vast majority of people with Barrett's never go on to develop cancer." He duly signed up and Reid's colleague Dr. Patty Blount performed the endoscopy. She divided the esophagus up into quadrants and took biopsy samples every centimeter in every quadrant of the affected region. "Each biopsy is about the size of a grain of rice they told me. Figure it out for yourself—thirty or forty grains of rice out of your esophagus every time. Boy, did I have a sore throat for days and days afterward! And until the pathology reports came back each time, I was a mess!" To his surprise—and relief—the Seattle team flatly contradicted his Michigan medical advice. The condition was low grade, and he was in no immediate danger. They have kept a watching brief ever since, and, so far, no cancer has developed.

This is the painstaking approach that Reid's team has taken to each member of their cohort every twelve months, steadily collecting data on the timing and frequency of mutations and the occurrence of chromosomal instabilities like tetraploidy and aneuploidy. It is beginning to pay off. They divided their cohort up into one group that did not progress to cancer and one that did, and compared the histology from years of endoscopic sampling in each of the patients to log the timing and type of genetic abnormality that had occurred. They have discovered a crucial prediction time window between four years and two years before esophageal cancer develops where critical differences in degree and type of abnormality occur in patients who progress to cancer compared to patients who do not. Prior to four years b.c. (before cancer), the esophageal lining of both groups was very similar, with chromosomal alterations including mutations in several genes that normally suppress tumor formation, DNA deletions, the loss of the small arm of chromosome 9, and increase in gene copy number on chromosomes 8 and 18. These abnormalities remained stable throughout the length of surveillance in the non-progressors, but in the group that eventually progressed to cancer, the genome rapidly became increasingly more disorganized and abnormal. "What we actually found was a sudden increase, it was as though the genome instability occurred very suddenly and very rapidly, and so the patients would undergo gains and losses of chromosome arms or whole chromosomes four years before the cancer and two years before the cancer they would undergo a catastrophic chromosome doubling. Those were the key events."

Reid has shown that rapid, large-scale genomic change—macromutation, if you like—is what distinguishes the non-progressors from the progressors in his study. He avoids referring to Goldschmidt because geneticists, who believe the "hopeful monster" idea was killed off in the 1940s, just stop listening. But Goldschmidt's description, he believes, accurately conveys what is going on. "Cancers are hopeful in the sense that they are just throwing up anything they can, and they are monsters in the sense that they can kill us. Although the vast majority of these 'monsters' are less fit than the predecessor cells, they are throwing them off so frequently that, once they go, they appear to be able to grow very fast. The frightening thing is we don't know how many times they can do that. To really control cancer we're going to somehow have to control this 'hopeful monster' process."

Bob Tell is typical of those non-progressors who develop some genomic abnormality but go no further, and he continues to have regular biopsy checkups in Michigan, which follow the Seattle protocol. For him, living with the continual uncertainty, the threat of cancer forever on the horizon, is far better than enduring life after drastic surgery. He recently heard from Patty Blount. "Patty said to me, 'Well, you're sixty-seven now, Bob, something's going to get you, but I'm now willing to bet it won't be your esophagus!' I would do anything for Brian Reid," says Bob, his voice choking with emotion. "He saved my esophagus."

Reid's research dramatically highlights the need for extensive, regular biopsy of Barrett's to identify the warning markers he has identified, if they occur. And this should apply to all potentially fatal cancers. But, in the wider oncology community, the penny has yet to drop. For instance, a recent survey of gastroenterological practice in the UK concluded that "90% of specialists did not take adequate biopsies for histological diagnosis. Furthermore, 74% would consider aggressive surgical resection for prevalent cases of high-grade dysplasia in Barrett's oesophagus as their first-line choice despite the associated peri-operative mortality."

As a cancer sufferer, you have a much better chance of staying alive if the tumor inside you stays in one place. Primary tumors seldom kill you. The problem is that most tumors eventually spread, or metastasize, to other organs. This is the endgame—cancer's last great throw of the dice—and it is invariably fatal. If tumors were homogeneous masses of cells, there would be no metastasis, but, as we've previously noted, tumors, particularly as they grow in size, are not homogeneous. They are miniature ecosystems with great internal diversity. Levels of nutrients and oxygen vary widely across the tumor mass, access to blood capillaries varies considerably, as does vulnerability to immune system attack. Reactive oxygen molecules constantly attack cancer cells. There is also vicious competition for resources between cancer clones, because as the tumor becomes more malignant, the cells inside it tend to grow and divide more rapidly, gobbling up to two hundred times the amount of glucose required by normal cells. The breakdown of sugars inside the cancer leads to buildup of acid, and this is also a strong promoter of the invasiveness and metastasis we associate with the later stages of malignancy. Oxygen gets rapidly depleted toward the center of tumors, and the resulting hypoxia is also a prime mover.

According to cancer researcher Athena Aktipis, cancer cells with their increased metabolism and rapid proliferation are resource needy and the equivalent of a greedy species that overgrazes its habitat in the world at large. According to this ecological theory of cancer, the greed of cancer cells is their Achilles' heel—it makes them vulnerable, and this creates the selection pressure for them to move on to pastures new. Staying put is no longer an option—they will starve to death. Metastatic cells, those that pack up and leave, are the cells that have lost out against these extraordinary levels of competition. But becoming a frontiersman comes with risks. Fortunately for us, the vast majority of the millions of metastasizing cells that leave a primary tumor every day never make it. As Carlo Maley puts it: "If you are a cell that is at a disadvantage in a tumor and this selects for your clonal progeny to leave the tumor, it is a very bad gamble because it is estimated that only one in a million cells succeeds and establishes somewhere else."

Before they can leave the primary tumor, cancer cells must undergo what is known as epithelial-mesenchymal transition in which the close-packed sheet-like structures we associate with the gut lining and lining of breast ducts, for instance, gives way to a much more loose-knit assembly of cells with much greater powers of migration. It is stimulated by hypoxia and involves the loss of the principle protein molecule, E-cadherin, that causes cells to stick together. Although tumors appear to be quite hard to the touch, individual cancer cells are remarkably soft, and cells destined to migrate particularly so. Metastatic cells are very good at burrowing through the extracellular matrix, the sticky, fibrous tissue in which all cells are embedded. They also deform, rather like amoebae, which allows them to shoulder their way through the matrix and squirrel in and out of pores in the walls of blood vessels, so that they can hitchhike in the bloodstream to the next capillary bed. This is why many secondaries are located in the lung—the initiating cancer cells get trapped there in the fine blood capillaries and begin to divide to form a new colony.

Why are metastases (they are known as mets in the business) invariably fatal for the patient? No one really knows, but there are theories. When cancer cells spread to other organs, explains Maley, they encounter different environments with different selection pressures. This means that each met will soon evolve different characteristics to any other met, and to the parent cancer. Any drug selection based on characteristics of the primary tumor will therefore be much less efficient at tackling the mets. Also, if the mets, on entering a new tissue, find they do not have the right growth factors or life support to hand, they will just sit there dormant. In this state they are almost impossible to kill because most chemo only targets proliferating cells.

Cancer clinicians often talk about their patients simply wasting away. They are referring to a condition called cachexia that contributes directly to about 20 percent of cancer deaths. It causes fatigue and breakdown of body tissues, particularly the muscles, and may be caused either by substances, as yet unidentified, from the cancers themselves, or through tissue damage caused by the massive immune response mounted at the different metastatic sites. As the body weakens, infections can eventually cause immune system collapse and so-called cytokine storms, unstoppable overproduction of inflammatory chemical signaling molecules, can occur, causing irreversible damage to vital organs, which soon fail.

Despite the upbeat messages from the cancer research charity community, the exhortation that "together we can beat cancer," the prognosis for the immediate future is sobering. There have been successes, like the huge improvement in survival rates in leukemia, and mortality from cancer has been slowly dropping every year in Western countries. But the number of cancer types that have increased their *incidence* in the last ten years vastly outweighs those whose rates have decreased. And while no one would disagree that healthy changes to diet, smoking, exercise levels, and body weight could reduce cancer incidence by about one-third across the board, the fact remains that we are estimated to have twice the cancer burden worldwide by 2030 than we have today. Early screening, vaccination against virus-caused cancers like HPV, and improvements in surgery, radiation, and chemotherapy have all played their part. But the truth of the matter is that the longer a cancer survives, the more complex it gets in structure, clonal diversity, and genetics, and the greater is the failure of targeted chemotherapy. And when a cancer metastasizes, there is very little any clinician can do.

One major drawback of much contemporary cancer chemotherapy is that it creates problems for itself because it selects for preexisting gene mutations within the tumor that confer resistance. Mel Greaves is astounded that the majority of clinicians have not yet realized, by comparison with antibiotic resistance—which is now of such proportions that it is threatening a world without effective antibiotic cover—the scale of the problem they are stacking up. "I don't think they think about it from an evolutionary point of view, even though we already have a prime example of what can happen with antibiotic resistance," he points out. "It's probably changing now a bit, but it's quite frightening that oncologists, up to now, have rarely if ever considered chemotherapy resistance in the same breath as antibiotic resistance. Extraordinary naïveté." Chemotherapy is still in the Stone Age. "Modern treatment is terribly crude; it's little more than what the ancient Greeks were up to, chucking poisons about. Most of the cytotoxic drugs will hit any rapidly dividing cells hard. You get transient benefits, because most cancer cells are dividing, but these cells can hunker down, just like a bacterium under stress will do, and if they are dormant, they are not dividing and therefore, by definition, they are resistant." Chemotherapy is also genotoxic; it can destabilize cancer cell genomes and help to produce the very mutations and genetic instabilities that make cancer malignant in the first place.

As we have seen, close examination of any maturing cancer reveals an extraordinary level of clonal diversity. "Clearing up such subclonal diversity is expensive, time-consuming, and at the end of the reach of contemporary technology," explain Greaves and Maley. "It represents a real challenge to completely curing cancer." Each cancer is unique, genetically diverse, preloaded with resistance, and prone to metastasis. In the vast majority of cases, all chemotherapy does is provide the selection pressure to push cancers toward malignancy, not away from it. And although pioneering research from scientists like Reid, Maley, Graham, Greaves, and their various colleagues holds out much hope for better prediction, allowing clinicians to discriminate between cancers likely to become malignant and cancers likely to remain benign, it doesn't solve the problem of what to do if the lesion does turn out to be malignant. All this has led some evolution in cancer researchers to suggest a revolutionary idea—if you can't beat it, why not live with it? They are experimenting with drug regimes aimed at stabilizing cancer rather than eradicating it.

At the Moffitt Cancer Center in Florida, Bob Gatenby, Ariosto Silva, and Bob Gillies have been mathematically modeling cancer to get a handle on why chemotherapy so often fails to eradicate it. Toward the center of a solid tumor, conditions become increasingly hypoxic due to lack of oxygen supply. They also become more acidic because the cancer cells have had to switch to anaerobic respiration. This leads to a core population of cancer cells that have become resistant to chemotherapy but have also had to shoulder the extra metabolic costs of surviving in this harsh environment. They are vulnerable to anything that further affects their metabolism. For several decades, clinicians have used a glycolysis inhibitor called 2-deoxyglucose (glycolysis is the chemical pathway that metabolizes glucose) to target these hypoxic, vulnerable cells. But they have used it simultaneously with chemotherapy. This doesn't have the desired effect because the chemotherapy selectively kills off the more nutrient-rich rapidly dividing cells at the outer rim of the tumor and allows glucose and oxygen to permeate toward the core, offsetting the starvation effect of the 2-deoxyglucose. Gatenby's group mathematically simulated a strategy that applied 2-deoxyglucose before chemotherapy. This produced a core of starving, dying chemo-resistant cells. They then modeled the pulverization of the exterior rim of the tumor with chemotherapy, and the simulation suggested that repetition of this treatment cycle would eventually lead to a tumor with no central chemo-resistant population. It was then potentially curable.

In a move to contain cancer rather than try to smash it into submission, they have come up with an idea called adaptive therapy by which chemo is not given at the same dose every day, but pulsed periodically. They have completed an exciting pilot project where they introduced tumors into mice. One group of mice then received regular doses of carboplatin at a rate of 180 mg/kg, while another group was given periodic adaptive therapy starting off with carboplatin at 320 mg/kg. The mice receiving the standard therapy initially responded well, but the tumor eventually recurred and they died. The adaptive therapy group received modulated doses that gradually decreased with time so that, by the end of the experiment, the mice proved able to survive indefinitely with a small, stable burden of tumors that were held in check by only 10 mg/kg of carboplatin.

Gillies, at the Moffitt, has tried another tack inspired by the knowledge that the hypoxic, acidic conditions in the center of tumors select for cancer cells that are particularly capable of metastasizing and invading other tissues. He reasoned that if he could raise the pH inside tumors—make them more alkaline—he might be able to reduce metastasis. He used mice with metastatic breast and prostate cancers and fed the mice a regular dilute dose of bicarbonate of soda in their drinking water. Although the bicarb had little or no effect on the size of the primary tumor, it significantly reduced the number and size of metastases to the lung, intestine, and diaphragm.

It is well known that inflammation in the tissue surrounding tumors stimulates them to become more genetically unstable and favors gene mutations that are associated with genetic instability and metastasis. Brian Reid has investigated the role of aspirin, a potent anti-inflammatory, and has shown that it reduces the risk of esophageal cancer. Other studies have shown the same effect for a variety of cancers. But aspirin goes further: "If you do a list search on cancer and genomic instability, you get thousands of hits," says Reid, "If you do a search on aspirin and cancer, you get thousands of hits; but if you do a search on aspirin and genomic instability, you can count the number of hits on your twenty digits! No one has said 'Hey—there is a link here!' But somehow aspirin is reducing genomic instability. It also decreases the production of new blood vessels in cancers, increases apoptosis, and reduces inflammation—that's surely a description of a miracle drug!"

No one yet knows how aspirin works, but there is a black irony here, and the joke is at the expense of the pharmaceutical industry, which spends and earns billions of dollars every year on chemotherapy and cancer immunology. It is that the most promising advice that has so far come out of several decades of research on the evolution of cancer is to use chemotherapy more sparingly to allow patients to live with their illness, rather than expect it to become eradicated. And the realization of the therapeutic promise of two of the most garden-variety chemicals you can imagine, that can both be bought for mere pennies—aspirin and bicarbonate!

Meanwhile, "hopeful monster" cancer research gathers pace, and more is being learned about the diabolical lengths to which cancer clones are capable of going in their march to malignancy and metastasis. So much research is now being done on the Darwinian evolution of cancer, that it is now possible to make generalizations about the mechanisms involved across a whole range of cancers. In the peculiar world of cancer, where clones of cancer cells battle and jostle for oxygen and nutrients, are besieged by the host's immune system, and blitzed and battered by chemotherapy, immense selection pressures are leveled at each cancer clone, which create profoundly aberrant cancer genomes in the desperate and rapid search for new sources of genetic variability to counteract them. Unfortunately for us, cancer is a niche in biology where Goldschmidt's idea of hopeful monsters still holds sway, and it is capable of producing malignant, metastatic tumors, which, for the most part, defy our means to eradicate them.

A PROBLEM WITH THE PLUMBING

WHY THE EVOLUTION OF CORONARY ARTERIES MAKES US PRONE TO HEART ATTACKS

When it comes to heart and artery disease, we live in a prevention-driven world with no shortage of do's and don'ts. Don't smoke; cut down on saturated fats; watch your weight; drink less booze; eat more fish, fruits, and vegetables; get more exercise. The advice is endless. Yet throughout the Western world, heart disease still remains the major killer. In the United States, over 1 million Americans will have a heart attack every year, and 600,000 will die. Heart disease costs American health care over a hundred billion dollars annually. In the UK, the statistics are just as dismal. Although incidence of heart attacks has halved over the last twenty years, it still accounts for one death every six minutes—46,000 of which are premature deaths that occur before we reach the proverbial three score years and ten. Thanks to an aging population and a cultural swing to fast food, Japan's heart attack rate is on the rise. All of us, if we have passed the age of adolescence, will have some sign of atherosclerosis in the walls of our arteries. We are all walking around with furred pipes.

Our coronary arteries are extremely vulnerable to disease. They are very small-bore vessels measuring only two to four millimeters across and so can easily be blocked by atheromatous plaque. Yet they perform the most vital job in the body because they supply the organ that supplies everything else with oxygenated blood and nutrients. They are the way the heart feeds itself. Ever since the origin of our vertebrate ancestors, the fishes, hundreds of millions of years ago, and the arrival of the first types of sophisticated hearts, evolution has been wrestling with the problem of how to supply blood to our most important muscle. This anatomical struggle is best framed by borrowing an idea from the famous "Who shaves the barber?" paradox, first coined by the philosopher Bertrand Russell. The barber shaves every other man in the village, but who, then, shaves the barber? The heart supplies blood to every other tissue and organ in the body, but how is the heart itself supplied with enough blood to allow it to work efficiently? As we shall see, evolution has ingeniously solved the problem in several ways in fish, amphibians, and reptiles, but the solution, the way the barber got shaved, in humans has inadvertently exposed us to the risk of heart attacks and premature death. The design of the blood supply to the human heart is a classic case of evolutionary compromise, but, equally, evolution is leading us toward a number of exciting medical technologies to repair damaged hearts caused by a problem with their plumbing.

Peter Berry, in his late seventies, has been living with heart disease for twenty-five years. He remembers that he and his wife had traveled on holiday to the seaside in 1986, which would have made him about fifty-three years old, and he had been hit by a severe heart attack only two hours after they arrived. He claims to have had absolutely no prior warning that his heart was developing problems. "I was sitting relaxing and I suddenly had terrific pains in the chest. . . . It just came on." He was a cable-jointer with the Electricity Board at the time, and although he kept himself fit, he did smoke. He was rushed to the hospital where a doctor said: "If I were you I'd give it up, it will kill you in the end!" They kept him in for a number of days while they performed an electrocardiogram, stabilized him, and put him on drugs.

Back home in north London, he was referred to his local hospital, and the doctor there told him: "Just carry on. We'll give you some more drugs to take. You've had a heart attack—you've had a warning." He rested up for six weeks before resuming light duties and, eventually, returned to full-time work. He soldiered on with regular hospital checkups to listen to his heart and do routine electrocardiograms, and he occasionally suffered minor angina spasms especially if he physically overexerted himself. Although he did not know it, his atherosclerosis, the gradual blockage of his arteries, was progressing silently but fast. Eventually the hospital gave him an angiogram (an X-ray image of the heart) and discovered that one of his coronary arteries was blocked. So, nine years after the initial heart attack, he had a stent inserted. A stent is a short length of wire-mesh tube that is pushed into the damaged part of the coronary artery to hold the lumen open. However, they were already too late to save the muscle on the left side of his heart—years of oxygen starvation had killed it off—so they satisfied themselves by placing a stent in another branch of one of his coronary arteries that was also showing advanced signs of atherosclerosis. He was by now sixty-two years old, and his employers were looking for workers willing to accept voluntary redundancy. He took it. "At least now we can be together and do what we want to do," he told his wife.

Consultant cardiologist Andrew Wragg sees plenty of patients like Peter Berry. A great deal of his routine work at the Barts Heart Center in London involves taking angiogram images of the heart and the blood vessels that supply it, to locate the exact position of clots in the coronary arteries. He then uses angioplasty, the technique that inserts stents in the coronary arteries, to reopen them. The worst kind of heart attack he sees is called a STEMI, where one branch of the coronary artery has become completely blocked. The heart is, essentially, an electrical organ. The beating of all the individual heart muscle cells is synchronized by a master pacemaker so that they can operate in concert to provide the powerful contractions the heart needs to pump blood around the body. Imagine the wristwatches of every person in New York to be literally enslaved to the Grand Central Station clock! Each heartbeat on an electrocardiogram shows up as a complex wave pattern, which is divided up into five segments, named after letters of the alphabet: P, Q, R, S, and T. The ST segment is when the muscle cells of the ventricles, having discharged, are repolarizing themselves in readiness for the next contraction. Immediately after full blockage of a coronary artery, all the heart muscle it serves begins to die and this massive cell death shows up as a characteristic elevation of the ST segment of the ECG, hence the acronym STEMI, ST segment elevation myocardial infarction.

When a heart attack victim is rushed to the hospital in an ambulance, they are usually given oxygen and aspirin, followed by a clot-busting agent and morphine. If the ECG they are given immediately on arrival shows the characteristics of a STEMI, there is absolutely no time to lose. The blockage in their coronary artery must be located fast and blood supply to the heart restored as quickly as possible. This is called reperfusion. The more heart muscle they can save, the better is the chance that the patient's heart can continue to function adequately.

The exact position of the blockage is found by injecting X-ray opaque dye into the artery. It is then usual to try to pass a fine wire through it. The wire is hydrophilic coated so that it doesn't stick. The technology to turn and steer this wire makes the process relatively easy. Equipment is then slid up the wire to suck out the clots. This is followed by an inflatable balloon, which opens and widens the artery where it meets the atherosclerotic plaque. This allows the surgeon to place the wire-mesh stent as a permanent internal scaffold. Although small, the stents are incredibly tough. Even when you have to push quite hard to get the stent in place, they rarely fall apart.

Once the stent is positioned correctly, blood-thinning drugs will keep further clots at bay and, in the long-term, cholesterol and blood pressure–lowering drugs will prevent most patients from having to return with further complications. Nevertheless, 30 percent of all the people who die following a heart attack will die at home before the emergency services can get there, and a further 5 percent will die quickly in the hospital despite this emergency intervention.

The heart and its coronary arteries seem almost perversely designed to restrict blood flow to heart muscle just when it needs it most. When the ventricle contracts during systole, to force blood into the aorta, the branches of the coronary arteries that ramify throughout the heart muscle become compressed and cannot fill with blood. Therefore the coronary arteries only receive blood when the heart relaxes in diastole. This quite literally puts pressure on the coronary arteries because as exercise rate goes up, the period the heart spends in diastole becomes shorter—filling time becomes shorter at exactly the time heart muscle oxygen demands are highest. This is why anything that reduces blood flow through the coronary arteries, like atherosclerosis, will tend to cause angina when people exert themselves.

More importantly, unlike most vascular networks in most organs, the coronary arteries and their branches are terminal—like cul-de-sacs in suburban road networks they are blind ends, there is no way out. This means that if there is a blockage in one branch of a coronary artery, there is no way for blood to circumnavigate it by taking a diversion. None of the tissue served by that branch of the coronary artery can avoid oxygen starvation. The problem for many patients is that not all heart attacks are a traumatic

STEMI. A non-STEMI attack can block a small branch of a coronary artery or partially block a main coronary artery, causing chronic oxygen starvation and resulting in gradual heart muscle damage. This may not be at all obvious to patient or doctor until it is too late to prevent a critical mass of heart muscle from being killed off.

This is precisely what happened to Duncan Chisholm. He had a suspected mild heart attack in 1983, when he was fifty-nine, for which he was given no medical intervention. But it caused him to give up smoking on the spot. In 2000 he was found to need a partial knee replacement, and during the routine preoperative checks, the anesthesiologist spotted something worrying on his ECG. It suggested that he had a blockage of the right coronary artery. He decided, nevertheless, that Duncan was fit enough to withstand the operation and it duly went ahead. However, soon afterward, Duncan began to suffer from angina, and his GP elected for further tests. An angiogram showed that in addition to the blockage in his right coronary artery, there was further sign of disease in the left. His consultant performed a triple coronary bypass operation in 2003. Although the operation was deemed successful, within four or five months he started hyperventilating and further tests showed that a massive 60 percent of his heart muscle was dead. Although he had never knowingly suffered a heart attack, there had been a gradual, almost silent, starvation of his heart.

The course of Peter Berry's heart disease followed a similar traumatic path and ended up in the same dismal place. "I woke up with terrific pains in my chest. The wife was fast asleep, I was sweating, and so I went downstairs and made myself a cup of tea. Then I thought, 'I'm not feeling any better,' so I went back upstairs and sat on the side of the bed and woke the wife up and said, 'You'll have to call an ambulance. I've got some pains and I don't feel well at all.'" He went back into the hospital and was told he really needed a heart transplant but couldn't be given one because they didn't think he would survive the operation. His heart was in a really bad way, and there was nothing they could do to repair it; he'd just have to live the rest of his life on drugs. "I was getting terribly out of breath—I couldn't walk fifty yards up the road. But my main concern was my wife . . ." His wife was becoming ill with dementia, and Peter, heart disease or not, saw it as his job to try to care for her at home as long as possible. "I thought, 'I've got to look after her—I've got to care for her'—and I did—for another seven years." He had to cope with all his wife's needs on 20 percent of a working heart. "I suppose if I'd pushed it I could have had more help, but I just didn't want it. My marriage vows were to be with her until we were separated, and I believed in those vows. And that was it."

Historically, doctors have pointed the finger at smoking, a couch-potato lifestyle, and an unhealthy diet rich in salt and saturated fats as important risk factors for coronary heart disease. However, a recent huge meta-analysis of dietary change and reduction of risk of heart attack found no effect. It is becoming clear that these factors are not enough to explain the epidemic proportions of heart and coronary artery disease, nor is their correction enough, necessarily, to protect us from furred-up arteries. Acknowledgment of this has led to a sea change in the conventional wisdom of what causes heart disease. Clues have been arriving from a number of directions that there may well be an important ingredient missing—the role of the immune system.

For example, in 2011 Staffan Ahnve, from the Karolinska Institutet in Sweden, published the results of an extraordinary long-term study. He had collected data on every Swedish resident born between 1955 and 1970, identified a subgroup who had either their tonsils or appendix removed before the age of twenty, and matched their health outcomes over the next quarter of a century against a control population who had neither surgical intervention. The data set was so large, he even managed to accumulate a substantial number of youngsters who had both tonsils and appendix removed. He calculated that childhood removal of the tonsils increased the risk of early acute myocardial infarction (heart attack) by 44 percent while removal of the appendix increased risk of heart attack later in life by 33 percent. The risk was even higher in those children who had lost both tonsils and appendix.

The tonsils and appendix are lymphoid organs that form important parts of the childhood immune system. Although that importance wanes after the age of twenty, the inescapable conclusion, although further research is always needed, is that their removal can compromise the proper development of the adult immune system, leaving the body more likely to develop illnesses in which a malfunctioning immune system is the culprit. Several studies, says Ahnve, suggest that Hodgkin's lymphoma is associated with appendectomy and/or tonsillectomy, and both operations seem to be a risk factor for the development of rheumatoid arthritis and Crohn's disease, two very severe autoimmune diseases.

Although Ahnve's work has been criticized for not taking into account possibly confounding variables, the suspicion that these childhood operations increase the risk of heart attack adds atherosclerosis to the long list of pathologies in which an abnormal immune system is involved. It suggests that atherosclerosis is an inflammatory illness very similar to the range of autoimmune diseases we discussed in the chapter "Absent Friends." The fact that several of these autoimmune diseases, most prominently rheumatoid arthritis and type 1 diabetes, are associated with increased risk of atherosclerosis tightens that link. Interestingly, removal of the spleen, another lymphoid organ, also accelerates atherosclerosis. The scientists are not yet sure exactly what is going on. It may be that the removal of these lymphoid organs reduces immunity to atherosclerosis; it may be that their removal weakens immune defenses against pathogens that, once established in the body, might set off an inflammatory chain of events in the arteries; it may be that their removal could lead to the development of an autoimmune disease with its attendant risks of atherosclerosis; or the fact that these organs had to be removed in the first place may point to some abnormality in the patients' immune systems. The fact that this association between tonsillectomy, appendectomy, and atherosclerosis is only found in people who underwent these operations before age twenty (the effect evaporates afterward) reduces or even eliminates other potentially confounding factors like the effects of smoking, lifestyle, or diet.

Dr. Maciej Tomaszewski and his colleagues at the University of Leicester's Department of Cardiovascular Science have recently reported the results of a study of over three thousand men that shows a strong genetic link between fathers and sons for factors that predispose to coronary artery disease. They homed in on the Y chromosome and discovered that 90 percent of British men have one of two major variants called haplogroup I and haplogroup R1b1b2. They have found that the risk of coronary artery disease in men who carry haplogroup I is 50 percent higher than in other men. This risk is completely independent of the other major risk factors for coronary artery disease—high blood pressure, high LDL cholesterol, and smoking. The researchers believe that this haplogroup has a direct influence on the immune system. Specifically, they show that a number of genes involved in the passage of white blood cells across the inner wall of the artery increase their activity, as do several genes involved in the production of pro-inflammatory chemicals called cytokines. Meanwhile, genes that moderate the immune response decrease their activity. When the immune system loses its ability to regulate the immune response, it can lead to a range of autoimmune diseases. The Leicester researchers specifically make the link, saying: "This conclusion implies that haplogroup I carriers might have chronic derangements in homoeostatic (balancing) mechanisms of adaptive immunity, possibly with heightened inflammation affecting the cardiovascular system. A similar mechanism has been well documented in other complex disorders, i.e., in inflammatory bowel disease, in which deficiencies in immunity status can lead to increased systemic inflammation."

Medical scientists now know a great deal about how the slow, silent destruction of the walls of our arteries comes about. The vast majority of us will have developed what are known as fatty streaks just under the internal lining of our arteries from late childhood onward. These are mainly formed from fat-laden immune cells called macrophages. In many people, these will disappear with time, but in others, fatty streaks will go on to form full-bodied atheromatous plaques, and once these lesions start to form, they develop gradually, often over decades, until one of two final traumatic events precipitates a heart attack. Göran Hansson, from the Karolinska Institutet in Sweden, paints a graphic picture of the development of atherosclerosis in gruesome detail. He points out that until a few years ago, it was accepted wisdom that atherosclerosis was merely the passive accumulation of cholesterol in the arterial wall and that if one could control high blood pressure and blood cholesterol, coronary artery disease could be made a thing of the past. We now have to accept that this is palpably not true. Coronary heart disease is quite clearly a more complex, chronic inflammatory process.

Normally, the interior lining of our arteries, the endothelium, is slippery. It is like Teflon—nothing sticks to it. However, several factors can reverse this situation and make the arterial endothelium secrete cell adhesion molecules that, as their name implies, make the endothelium sticky. Here's where one of the major risk factors for heart disease, smoking, comes in. Nicotine and carbon monoxide in tobacco smoke damage the endothelium. Smoking also lowers circulating levels of high-density lipoproteins (HDLs), or "good" cholesterol. Normally, "good" cholesterol mops up the low-density lipoproteins (LDLs), or "bad" cholesterol, which are the real villains of the piece, and transports them to the liver, where they can be broken down. Thus smoking skews the ratio between good and bad cholesterol in the blood in favor of bad cholesterol and chemically changes the bad cholesterol to make it more atherogenic. Damage to the interior lining of the artery walls allows circulating LDLs to stick to them and then enter. Smoking also exacerbates high blood pressure (hypertension), which itself is associated with damage to the arterial lining. Blood flows in different patterns throughout the arterial system. Near branches in the arterial tree and on the inner wall of curvatures in the arteries, the flow can become disturbed and turbulent. This can set up shear stresses on the artery wall that switch on genes in the cells of the endothelium to produce adhesion molecules and cytokines, which can promote inflammation. Now the scene is set for the atherosclerotic plaque to grow.

Platelets are the first blood cells to arrive and anchor themselves at the site where this sticky change in the endothelium is taking place. They produce glycoproteins that stimulate other platelets to stick to them, forming a clot. Interestingly, platelets are only found in mammals, which, together with birds, are the only animals to have a high-pressure systemic blood system to take blood away from the heart and push it around the body. According to researchers at the University of Pennsylvania, headed by Mark Kahn, platelets may have evolved in mammals as a super-efficient clotting mechanism capable of stemming traumatic wounds to these high-pressure arteries. The downside, they say, is that cardiovascular disease may be the price we have to pay for such a protective innovation, when platelets aggregate only too easily around the roughened surface of arterial walls.

The adhesion molecules pouring out of the endothelial cells also stop white blood cells from rolling past and cause them to stick. Once anchored, more chemical messengers, called chemokines, cause them to migrate into the arterial wall, where they come to rest in the underlying layer—the intima. Here a cytokine called macrophage colony-stimulating factor causes these white blood cells to develop into macrophages, which are scavenging cells that engulf and destroy all sorts of rubbish like bacterial endotoxins and fragments of dying cells. This is a key transformation that underlies coronary artery disease and is precisely the step that the Leicester researchers found to be exaggerated in Y-chromosome haplogroup I men. The macrophages also take up the oxidized form of LDL, which has also been pouring through the endothelium into the space beneath. The engorged macrophages produce a further cascade of inflammatory cytokines, proteases, and free radicals, and become transformed into foam cells, the building blocks of full-fledged atherosclerotic plaque. The immune changes in the artery wall are now becoming pathological. Patrolling lymphocytes, known as helper T cells (so-called because they originate in the thymus gland above the heart and assist, or help, other immune factors to do their work), are attracted to the differentiating macrophages by the antigens on their surface. Behaving as if they had encountered a major site of infection, they secrete yet more types of cytokine, all of which skew the immune response toward inflammation.

Atherosclerosis now enters its acute phase. The atherosclerotic plaque bulges into the lumen of the artery. Inside it is a mess of T cells, mast cells, foam cells, dead and dying cells, and cholesterol crystals forming a necrotic core seething with pro-inflammatory cytokine molecules. All this is held together, up to a point, by a fibrous cap. The endgame can now take one of two paths: As the plaque continues to grow, it bulges out farther and farther. If this occurs in one of the coronary arteries, blood flow to the heart starts to become compromised, starving heart muscle cells of oxygen. This can lead remorselessly to angina. Occasionally the fibrous cap holding the plaque together ruptures, and the contents of the plaque spill out into the lumen of the artery, where they attract further platelets to the aggravated site. Blood coagulates and the resulting thrombus blocks the lumen of the artery completely or is carried, as a floating embolus, farther down the artery to a point where it can lodge, blocking further flow of blood. This is a heart attack.

If you ever find yourself lying in a hospital emergency department, frightened and in pain, head swimming with morphine, with electrocardiogram terminals taped to your chest, you might be forgiven for reflecting ruefully on just why the most important muscle in your whole body, the one that supplies everything else with oxygen and nutrients, depends for its own blood supply on two narrow pipes, our coronary arteries, and their branches, which appear to be so prone to fatal or near-fatal blockage. Why has evolution designed us with such easily compromised plumbing?

The short answer to this is that it hasn't. Despite all the recent research on the immune component of heart disease, one popular evolutionary explanation still places the blame for coronary heart disease on us, and not some inherent design fault in our plumbing. This theory claims that coronary artery blood supply was a perfectly efficient, adequate, and dependable way of feeding the heart until the last hundred years or so when the modern Western lifestyle caught up with us through a lethal combination of smoking, diets rich in saturated fat, lack of exercise, and occupational stress. The idea is called "mismatch." It maintains that our diet and lifestyle are dangerously out of kilter with the Paleolithic hunter-gatherer-type lifestyle that typified much of human evolution. Heart disease, according to mismatch theory, is a disease we have brought down on our own heads in modern times. But mismatch theory is now under challenge thanks in part to a controversy inside the multidisciplinary science of paleopathology over how best to interpret the evidence for the past relationship between human diet and heart disease.

Forty years ago, Rosalie David, who was then the keeper of Egyptology at the Manchester Museum, pulled together a number of scientific specialists in radiology, computed tomography, and other types of non-invasive medical imaging to form what became known as the Manchester Mummy Project. They gradually accumulated a collection of tissue samples from Egyptian mummies held in collections all over the world. They've been investigating the anatomy of mummies ever since, have recorded widespread heart and artery disease among ancient Egyptians, and collected evidence from a plethora of ancient texts that they believe links heart disease and lifestyle. They weren't the first: calcification (atheromatous plaque) of the aorta of an Egyptian mummy was first noticed by Johann Czernak in 1852, and the Anglo-German pioneer of paleopathology, Sir Marc Armand Ruffer, identified arterial lesions in hundreds of mummies dating between 1580 BC and AD 527 while working as professor of bacteriology at the Cairo Medical School at the beginning of the twentieth century. One anatomical section made by Ruffer shows a huge atheromatous lesion completely blocking the subclavian artery. Atheroma was identified in the arteries of King Menephta, Ramses II, III, V, and VI, Sethos I, and many others, by these paleopathology pioneers.

David reported in the *Lancet* in 2010 that her group had identified sixteen Egyptian mummies where sufficient remnants of hearts and arteries remained to permit analysis. Significant calcification was seen in nine of them. Meanwhile, a joint US-Egyptian research team co-headed by Dr. Gregory Thomas of the University of California, Irvine, and Dr. Adel Allam, professor of cardiology at Cairo's Al-Azhar University, has performed CT scans on fifty-two mummies housed in the Cairo Museum. Over half of them showed evidence of calcification of the arteries. The team also documented the earliest case of atherosclerosis of the coronary arteries ever recorded, that of Princess Ahmose Meyret Amon, who was born in 1580 BC and died in her forties. All branches of the coronary artery supply to her heart were blocked.

What factors might have caused this epidemic of heart and artery disease in ancient Egypt? Here is where the disagreement begins. David and her colleagues are in no doubt that they are the same factors they believe cause atherosclerosis today, particularly a diet rich in saturated fats. Remember, David points out, that only royal and other high-status families would have been embalmed and mummified, and there is ample evidence that they enjoyed a much richer diet than the hoi polloi. David's group has translated hieroglyphics taken from temple walls that give details of the daily food offerings to the gods. Priests and their families subsequently ate this food, she says, and so it gives a very good indication of the dietary habits of the upper classes. The diet consisted mainly of beef, wildfowl, bread, fruit, vegetables, cake, wine, and beer. Many items would have contained saturated fats. Goose, she notes, which was commonly consumed, contains around 63 percent of its energy as fats, 20 percent of which would have been saturated, while Egyptian bread was enriched with fat, milk, and eggs. Ancient Egyptian intake of saturated fats, explains David, probably significantly exceeded dietary guidelines for saturated fat that exist today; their food was also high in salt, which was used as a preservative, and they also liked to drink. David cites the case history of the Leeds Mummy, which her group examined. The mummy was that of a priest and had been brought to the Leeds Museum in the nineteenth century. He had died in middle age, and inscriptions buried with him established that he regularly consumed food offerings largely consisting of cuts from sacred cattle, kept in the temple precinct for exclusive use as offerings. When tissue samples were taken from his femoral artery, they showed well-developed atheromatous plaques.

David is unequivocal. In the modern epidemics of heart and artery disease, we are simply revisiting the sins of the past. Heart disease is, and always was, hastened by imprudent diet. Gregory Thomas and his team beg to differ. They have no quarrel with the ancient papyri that chronicle dining tables groaning with honey cakes and fatty meat. However, they note, several of the prominent risk factors today, like lack of exercise and smoking, were not shared by the ancient Egyptians. They do not seem to have been, as a rule, obese, and they were not couch potatoes. However they would have been exposed to high and chronic levels of pathogens, and no stratum of society would have been immune. Malaria and schistosomiasis would have been rife. It is possible that their immune systems were in a permanent state of inflammatory response in the attempt to repel these pathogens, and, as we have seen, an immune system skewed toward inflammation is a definite risk factor for the development of atherosclerosis.

Further challenge to David's conclusions comes from a glacier on the Austrian-Italian border, where Ötzi the Iceman was discovered in 1991. He had died in the Austrian Alps over five thousand years ago and had remained preserved in ice ever since. A tear in an artery wall had finally put paid to him, and his short, hard thirty-five to forty years of life had left indelible scars on his body. There were arrow wounds, and he had suffered from severe osteoarthritis. His last meal had consisted of mountain deer and cereals, eaten only hours before he met his death. Painstaking tomography revealed teeth full of cavities. His advanced dental caries was likely caused by a diet rich in carbohydrates and had exposed him to chronic bacterial infection, another risk factor for heart disease because, again, it leads to a chronic inflammatory immune response. His arteries were riddled with atheromatous plaque. Very recently, scientists from the Institute for Mummies and the Iceman, in Italy, have retrieved enough DNA from Ötzi to complete a comprehensive read-out of his entire genome. They have discovered that it contains certain gene variants that would have greatly increased his risk of developing heart disease.

Now Gregory Thomas and his colleagues Adel Allam and Michael Miyamoto have joined forces with a large multidisciplinary team headed by Randall Thompson of Saint Luke's Mid-America Heart Institute in Kansas City. They have collected 137 whole-body computed tomography scans of mummies from four very different geographical regions: ancient Egypt, ancient Peru, ancient Pueblo Indians from the southwest United States, and the Unangan of the Aleutian Islands in the far northern Pacific. Atherosclerosis was widespread in all these ancient civilizations.

While noting Rosalie David's claims for the very fatty diets of high-caste ancient Egyptians, these researchers point out that the diets of the other three civilizations were very different and disparate. The Peruvians farmed corn, potatoes, manioc, beans, and hot peppers, and they obtained much of their protein from alpaca, ducks, wild deer, birds, and crayfish; the Pueblo Indians were forage farmers who subsisted on maize, squash, pine nuts, and grass seeds, and relied for their protein on hunted rabbits, mice, fish, and deer; the Aleutians collected berries on land and obtained pretty much everything else from marine animals like seal, sea lion, whales, and assorted shellfish. Most of them ate what would be called a prudent diet today, or a diet heavily laden with unsaturated fat; none of them could be called remotely sedentary.

The team did note that the Pueblans and Aleutians used fire extensively for cooking in closed surroundings, with evidence for smoke inhalation in their lungs. This could have played a part in the development of their atherosclerosis, much as cigarette smoking is suspected of doing today. But, for them, far more important was the very huge infectious load all these ancient civilizations would have carried. In twentieth-century hunter-forager-horticulturalists, they note, 75 percent of the mortality was due to infections, against only 10 percent from old age. They conclude that the high level of chronic infection in premodern conditions might have been as inflammatory as in the accelerated atherosclerosis experienced today by patients with rheumatoid arthritis and systemic lupus erythematosus—a whole-body autoimmune disease that can affect heart, joints, liver, kidneys, and the nervous system. They are hoping to analyze DNA taken from the mummies and speculate that they might find gene variants that evolved to produce robust, pro-inflammatory immune systems to fight off all these infections in early life, at the expense of widespread, inflammation-induced heart and artery disease later on.

It looks as if atherosclerosis may be a disease of an unregulated and skewed immune system that has always been a feature of human pathology, even in ages less prone to dietary excess, lack of exercise, and the abuse of our bodies with tobacco and alcohol. It is a pathological process to which our coronary arteries have always been particularly susceptible. In order to understand why coronary arteries arose in the first place, we need a brief guided tour of the vertebrate kingdom to explain the several different methods that evolution has contrived to get an adequate blood supply into muscular hearts.

Colleen Farmer, a professor of biology at the University of Utah, has charted the way that evolution has coped with the immense problem of oxygen supply to the heart from the ancestors of the vertebrates in the pre-Cambrian period, 500 million years ago, to higher vertebrates like us. Our first ancestors, she explains, were marine suspension feeders who had beds of cilia that were used to trap food, where, later, fish had gills. Respiration was carried out through the skin. We have a living example today of these early ancestors in the primitive larval lamprey. Its simple heart pumps blood over the cilia in its throat, and the nutrient-rich blood flows round the body to the skin, where carbon dioxide is off-loaded and oxygen taken up, before it returns to the heart. So, because the heart is downstream of the main organ for oxygen exchange—the skin—it will always receive a blood supply fairly rich in oxygen.

As the bony fishes evolved, however, gills lost their function as filter-feeding mechanisms and adapted for gas exchange. Now the heart was upstream of the source of fresh oxygen. It pumped stale blood over the gills, which replenished the blood with oxygen and then propelled it onward to all the body tissues. But by the time the blood returned to the heart, it was severely depleted in oxygen and there was little left to sustain the heart. This may not have been unduly limiting in those fish that tended to have very inactive lifestyles, but many bony fishes were evolving into highly active predators. Their blood-circulation arrangements were severely cramping their style. The more active gill-breathers become, the more oxygen their skeletal muscles withdraw from the blood supply, and the more deoxygenated is the blood returning to feed the heart just when it needs oxygen the most. Worse still, the metabolic by-products of respiration enter the blood, turning it acidic. This acidosis decreases the ability of hemoglobin in the blood to pick up oxygen at the gills. The overall effect can be fatal. Farmer notes that these lethal effects of intense exercise are still commonly seen today in a variety of gill-breathing fish, including cod, dab, tench, salmon, bass, haddock, and trout. It may be that this is why some fish caught in fishing tournaments fail to survive even though they are uninjured except for the exercise. Oxygen starvation of the heart would have set up strong selection pressure for the evolution of a means to alleviate it.

Farmer believes this resulted in the evolution of lungs in many fish species. Conventional wisdom, she says, has it that lungs arose as a special adaptation in gill-breathing fish that inhabited shallow seas, brackish estuaries, or freshwater swamps where oxygen was frequently in poor supply. Lungs allowed them to breach the surface and gulp in air. An extension of these conventional ideas suggests that the evolution of lungs for this reason became an initial stage in the migration to land of the early amphibians. Farmer disputes this by pointing out that the groups of living fish that provide the best analogy for the last common ancestor of the two lineages of bony fish, the sarcopterygians and the actinopterygians, have lungs. This common ancestor lived at least 50 million years before the first amphibians. Furthermore, as the fossil record has increased over the years and our understanding of the habitat of these animals has improved, it has become clear that these early fish would have inhabited well-oxygenated seawater in wide-open seas and oceans. Lungs were therefore *inherited* by the lineages of bony fish now living in freshwater swamps, and not evolved to adapt to anoxic conditions. They were an ancestral, not a recently evolved trait. Furthermore, the contemporary environments of the Australian lungfish and other lung-breathers like the gar and the bowfin have been shown to be far richer in oxygen than previously implied, suggesting brackish water was not the prime selection pressure for lungs.

If you look at the anatomy of the circulatory system of a typical lungfish, it appears at first sight to be very inefficient. Blood is first pumped from the heart to the gills, where carbon dioxide can be gotten rid of and oxygen taken up. Some of the blood is then pumped around the body to the tissues, but some is diverted to the lung. This blood, instead of going straight on to feed the tissues, immediately returns to the heart, where it is mixed en route with deoxygenated venous blood returning from the rest of the body. The cycle is then repeated. This means that systemic blood, circulating to the body tissues, is always a diluted mixture of oxygenated and deoxygenated blood. This apparent inefficiency only makes sense, explains Farmer, if lung circulation specifically evolved to feed the heart, not the body. Farmer backs up her highly unconventional idea by examining the environments and lifestyles of a host of lung-breathing fish and finds that the presence of lungs does not correlate at all well with life in shallow, brackish, oxygen-poor water. It has much more to do with supporting energetic exercise. For instance, the gar, once thought, she notes, to be a "sit-and-wait" predator, is in reality a highly active migratory fish, while two other lung-breathers, bowfin and tarpon, are highly sought-after game fish. The bowfin, she says, has been described as "one of the hardest fighters that ever took the hook."

Why, then, don't all fish have lungs? Most living fish today are teleosts—bony fishes—that have converted their lung into a swim bladder, a buoyancy device that allows stable swimming at any level. The other major group of fish that are alive today are the sharks. They do not have swim bladders but rely on their huge livers stuffed with lighter-than-water squalene, a relatively light cartilaginous skeleton, and dynamic lift caused by constant swimming to maintain buoyancy. If they stop swimming, they tend to sink.

Without a lung, most extant bony fishes and sharks are thrown back onto gill breathing, and yet many species have become extremely active over evolutionary time. Importantly, in the lamprey and in many species of fish (and in the amphibia and many reptiles), heart muscle, known as myocardium, differs greatly in its anatomy to the heart muscle of the higher vertebrates. Instead of the hard, muscular mass that comprises mammalian hearts, called compact myocardium, the hearts of most lower vertebrates are composed of a loose-knit mesh or lattice of heart muscle cells, open to the main chamber of the heart, called spongy myocardium. Blood entering the chambers of the heart can therefore flow freely through channels called lacunae in the spongy myocardium, supplying the muscle cells with oxygen. But there's a catch. The spongy heart muscle so typical of many more sedentary fish, and so easy to supply with oxygen from inside the lumen of the heart, simply cannot provide the strong muscular contractions needed for blood circulation to power a more active lifestyle. The more active the species becomes, the more the heart muscle has had to compact. For instance, compact myocardium forms about 10 percent of heart muscle in the dogfish, but between 30 and 40 percent in the trout, and a massive 60 percent in the highly athletic tuna.

Typically, the muscle toward the center of the heart, closest to the ventricular chamber, will remain spongy and be fed by direct transfer of oxygen from the blood passing through it. But the dense myocardium to the periphery of the heart loses its connection to ventricular blood. In these species we see the evolution of a system of coronary arteries to feed the heart from the outside. The coronary arteries of fish differ from those of land vertebrates and mammals in that they branch off the blood vessels that serve the gills. They augment rather than replace the supply of oxygen. Only for extremely active fish like tuna and salmon does a coronary artery circulation seem vital, and, interestingly, the migrating Atlantic and Pacific salmon, may, like us, suffer from atherosclerosis! Tony Farrell of Simon Fraser University in Canada has reported severe lesions in the main coronary artery of salmon. The lesions develop in young fish and progress with age. Farrell suggests that the initial damage is caused by physical distention of the artery during strenuous swimming. It could well limit blood supply to the heart, especially when the fish are forced to swim upstream to their spawning grounds through water that can become increasingly deoxygenated.

Amphibians have hearts that somewhat resemble—in outline—the pictorial cartoon version of the heart you find on Valentine's Day cards. The heart has two atria—right and left—sitting above one single ventricle. The heart muscle is composed of spongy myocardium. Deoxygenated blood from the body enters the heart via the right atrium, and oxygenated blood from the lungs enters via the left. But because there is only one ventricle, it means that both oxygenated and deoxygenated blood is pushed into it at the same time. The deoxygenated blood keeps to the right-hand side of the ventricle, and the oxygenated blood, while sharing the same ventricle, keeps to the left. The two bloodstreams—one deoxygenated, one oxygenated—move like well-behaved traffic on the highway. They keep rigidly in separate lanes and do not mix. They exit the ventricle in parallel before the deoxygenated blood peels off toward the lungs in a pulmonary vessel and the oxygenated blood peels off to the rest of the body through an aorta. This means that the spongy myocardium on the right side of the heart only receives poorly oxygenated blood. It seems that amphibians compensate for this, especially at times when they need to be active, by taking up oxygen from the skin, somewhat enriching the returning circulation to the heart.

In all reptiles, except crocodilians and birds, the heart is also three-chambered. It has left and right atria and a single ventricle with two partial internal walls or septa. If there were no mixing of oxygen-rich and oxygen-poor blood within the heart, the muscle on the right side of the heart would be permanently deprived of oxygen-rich blood from the lungs and resigned to receiving only oxygen-poor blood in the venous return to the heart. And, indeed, despite the fact that the internal walls of the heart are incomplete, the reptile heart manages, in the main, to keep apart the deoxygenated circulation on the right side of the heart, and the oxygenated circulation on the left, so the fact that there is a gap there at all has always been viewed rather sniffily by comparative anatomists as a primitive feature. A staging post, if you like, on the evolutionary road to the more efficient mammalian design. But Farmer would have it otherwise. During those periods when reptiles need to be active, and when the right-hand side of the heart would therefore quickly become stressed, they have the ability to shunt oxygen-rich blood through the gap in the walls between the ventricles from left to right, bathing the spongy myocardium of the right ventricle in oxygen-rich blood. Although this inevitably deprives their bodies of some vital oxygen, it does keep their hearts going. This feature is not so much a primitive evolutionary loose end as an adaptation that has served the reptiles well for hundreds of millions of years.

With the evolution of the crocodiles, birds, and mammals comes the full-fledged four-chambered heart. By entirely separating the pulmonary circulation from the systemic circulation, it ensures that the systemic circulation, leaving the heart via the aorta, can operate at much higher pressures to satisfy the increasing energy demands of these highly active animals. This allows the pulmonary circulation to operate at lower pressure to avoid damaging the fine capillaries of the lungs. But the price to pay is that the right side of the heart has no access whatsoever to oxygenated blood. Worse still, all the myocardium is now of the compact type that is impermeable to the blood circulating through the heart's chambers. Although the heart is now a powerful and efficient organ, heart muscle is now completely dependent on blood piped in from outside the heart via a network of coronary arteries.

Cardiology research is using insights from the evolution of vertebrate blood circulation, the evolution of the immune system, and developmental biology to provide us with new medical technologies to repair damaged hearts. They range from the ridiculous to the sublime. In the 1960s, the Indian cardiac surgeon Profulla Kumar Sen pioneered a technique to restore blood supply to heart muscle through tiny channels made by puncturing the heart wall with acupuncture needles. His inspiration came from dissecting the hearts of Russell's vipers and showing that the vast majority of the spongy myocardium was bathed in blood via the network of channels or lacunae ramifying through it. Sen actually called his procedure the "snake-heart operation." His idea was taken up and modernized by American cardiologists, including the famous Denton Cooley at the Texas Heart Institute. These surgeons punched a series of holes right through the damaged areas of myocardium using an 800-watt carbon dioxide laser. The holes soon healed on the outside of the heart leaving a number of channels leading into the heart muscle from the ventricular lumen, through which blood could force itself under ventricular pressure. Several trials suggested that patients obtained some relief from angina as a result of the operation—possibly through the placebo effect. But the precise way in which the technique improves blood flow throughout the muscle—if it does—is unknown. Not surprisingly, the technique has fallen out of favor.

More recently, and sensibly, medical scientists have reasoned that if atherosclerosis is basically an immune phenomenon, then it ought to be possible to produce some kind of vaccine against it. At the heart of these developments is Göran Hansson. Vaccines work by recognizing foreign proteins called antigens that are displayed on the coats of viruses. During the progression of atherosclerosis, T cells in the immune system recognize antigens presented by "bad" cholesterol, LDL, and attack it. This is what leads to the out-of-control inflammatory response. As long as the LDL keeps to blood, liver, and lymph, the T cells are inhibited from this overreaction, but once it crosses the arterial endothelium and enters the artery wall, it appears out of context and the T cells let rip. Hansson and his colleagues realized that if they could block, or vaccinate against, the receptor molecule that the T cells use to recognize LDL, they might be able to stop the runaway immune overreaction. In early experiments using mice in which they have induced atherosclerosis, they have reduced the disease by 70 percent. They are now devising ways of adapting the therapy for human use.

We saw in "Absent Friends" that there exists a subpopulation of T cells called regulatory T cells, or Tregs, which can tone down the immune response. This is the mechanism that the "friendly" bacteria in our guts stimulate to prevent the host immune system from reacting against them. By regulating the immune response, they are, in effect, passing themselves off as "self." These Tregs help to prevent the wide range of allergic and autoimmune diseases that afflict us today by inhibiting the runaway production of the helper T 1 and helper T 2 cells that cause them. Now Hansson and colleagues at the European Vascular Genomics Network in Paris have shown that in the same way that Tregs inhibit allergic and autoimmune diseases, they can also inhibit the growth of atherosclerotic plaques. They injected Treg cells into induced atherosclerotic lesions in mice and saw that they were able to stop atherosclerosis in its tracks, but when they added antibodies that blocked the action of the Tregs, the mice developed even bigger lesions.

The most vocal and successful proponent of the role of inflammation in heart disease is without doubt Paul Ridker, of Harvard Medical School. Ridker points out that there was a great deal of medical interest in the role of inflammation over fifty years ago, but that it became completely eclipsed by the cholesterol hypothesis and research into the role of blood lipids, particularly bad cholesterol, LDL. Now inflammation is set to make a startling comeback that offers the promise of a powerful predictive medical approach to heart disease. Ridker's starting point was the observation that too many stroke and heart attack victims had normal cholesterol levels for cholesterol to be the overwhelming risk factor, and his work over the last two decades has been to find more potent markers and predictors for heart disease.

Ridker has shown that all the pro-inflammatory cytokines that are implicated in autoimmune and allergic diseases, like interleukin-6 and tumor necrosis factor-alpha, are elevated in patients at risk for coronary heart disease. However, these are difficult and expensive to isolate and analyze in the lab. What was needed was a marker molecule that accurately predicted risk and could be simply measured in any standard blood test. He found it in C-reactive protein. Thanks to Ridker and his colleagues, we now know that CRP predicts risk of heart attack much better than LDL, and predicts future stroke or heart attack in men with no history of heart disease and in apparently healthy middle-aged women. Statins are widely prescribed to lower blood lipid levels, but Ridker has shown they are also powerful anti-inflammatory agents that reduce CRP in the blood. However, it has not been conclusively proved that consistently lowering CRP saves lives, and the hypothesis is currently being tested in two mammoth trials in the United States that are due to report in around 2018.

For hundreds of thousands of us every year, however, cures for the atherosclerosis clogging our arteries or warnings of the risk we are running will come too late. A heart attack has already happened, and widespread damage to the heart has been done. Our hearts contain approximately 4 billion cardiac muscle cells, called cardiomyocytes, but a severe heart attack can kill at least a quarter of these within hours. For many patients, the scarring of heart tissue, the large mass of dead or dying cardiac muscle, called the infarct, and gradual ballooning of the weakened walls of the heart represent a slippery slope to total heart failure despite emergency medical intervention. Heart transplantation is the obvious solution for these distressing cases, but there will never be enough donated hearts to go around. This is why, over the last decade, large groups of scientists in the field of regenerative medicine have turned to a variety of stem cell technologies in the hope that they will lead to therapy to replace dead cardiomyocytes with new. The ultimate ambition of regenerative heart medicine is to find a type of stem cell that can be injected directly into the damaged heart, either when the patient is in intensive care immediately following a heart attack, or much later when a patient has been living with a declining heart for many years.

According to Joshua Hare, professor of cardiology at the University of Miami, the researchers broadly divide into two camps largely depending on whether they take an optimistic view or a dim view of the heart's ability to repair itself by regenerating new heart cells. The former camp are looking at ways of coaxing the heart into mending itself, while the other camp assume that it does not have enough self-regenerative power and that some totally external stem cell technology has to be applied.

The emblem of the British Heart Foundation, by far the UK's biggest funder of cardiology research, is the zebra fish because it has been shown to spontaneously regenerate heart muscle cells after catastrophic experimental damage to its heart. The experimenters removed 20 percent of the ventricle of the zebra fish heart by snipping away its apex. A clot made up of fibrin and collagen soon formed over the amputated area, and, within days, the researchers noticed signs of widespread regeneration of cardiomyocytes throughout the heart. The injury had not caused some resident pool of undifferentiated stem cells to suddenly become active but had caused a significant number of mature cardiomyocytes to de-differentiate and return to an immature state. They first detached from each other, and then the muscular contractile elements inside the cells, the sarcomeres, broke down. They then started dividing to produce several generations of daughter cells, which then rapidly matured into a new population of heart muscle cells. This regeneration process has not evolved specifically to repair catastrophic damage to the heart, of course. Zebra fish, in common with newts and salamanders, have widespread powers of regeneration.

Higher vertebrates have largely lost this ability. If a day-old mouse suffers damage to its heart, it can regenerate itself, but only in a tiny time window of approximately seven days. After this, scar tissue forms and the damage is permanent. Why can't humans "do the zebra fish"? Scientists are still unraveling the mechanisms of tissue regeneration and have discovered a number of developmental pathways that have fallen silent in higher vertebrates, probably for good evolutionary reasons. It may be that longer-living animals like us face too much danger from cancer formation if our cells retain the ability to divide indefinitely.

For many years, the human heart was considered a "post-mitotic" organ because it couldn't replenish old or damaged cells by cell division. But recently it has been shown that there is some turnover in cardiomyocytes after all. Up to 50 percent—2 billion—of them will be replaced during our lifetime. Present techniques cannot decide whether this occurs by regeneration of mature cardiomyocytes or whether there exists a latent pool of heart muscle progenitor cells waiting to be activated, although most researchers favor the latter. The big question is whether that modest rate of cardiomyocyte renewal, spanning a lifetime, could be of any use in a heart attack emergency where up to 2 billion heart muscle cells might need to be replaced as soon as possible.

On the assumption that any modest regeneration would be swamped by a heart attack, several groups of researchers are trying to devise clever ways of making functional cardiomyocytes from completely different sources. In a perfect world, the automatic choice would be embryonic stem cells because these cells are genuinely totipotent and can be made to differentiate into any cell type in the body. However, there are several major problems to overcome before they can be used. Embryonic stem cells must be harvested from a day-five human embryo, and this raises ethical and religious concern in many parts of the world, especially the United States. Also, because they don't originate in the patient's own body, there will be an immune reaction to them, as there is to any transplant, which has to be overcome by suppressing the patient's immune system. Third, although progress is being made here, embryonic stem cells are notoriously difficult to keep "on message." It is difficult to culture them for the many generations needed to bulk up cell number while keeping them on the intended path to one specialized differentiated cell type. They also have an alarming tendency, once injected into host tissue, to form teratomas. These are peculiar encapsulated tumors that, although usually benign, become a nuisance because they may contain unwanted inclusions of eyes, teeth, or parts of other organs representing the multiple possible developmental fates of embryonic cells.

Because of these limitations, researchers have turned to other types of stem cells from the patient's own body. Work is progressing on so-called inducible pluripotent stem cells (iPSCs). These can be obtained from the lining of the mouth or even from a few plucked hairs, and induced, or reprogrammed, into a pluripotent state (capable of a more limited repertoire of cell outcomes than totipotent embryonic stem cells) by injecting them with certain genes. Research in vitro and in animal models has shown that these cells can be made to develop into functioning cardiomyocytes. For instance, Lior Gepstein, at the Technion–Israel Institute of Technology, and his colleagues have harvested a type of skin cell called a fibroblast from human heart failure patients and inserted three genes, called transcription factors, into them using a virus to deliver them into the fibroblast nuclei. This reprogrammed the cells, which were then induced to differentiate into cardiomyocytes. The researchers then put the new cardiomyocytes together with newborn rat heart cells and showed that the cardiomyocytes were able to integrate with the rat cells and synchronize their electrical activity with them so that they all contracted in unison. All this, of course, has only been done in vitro in the laboratory, and it may be many years before it can lead to human clinical trials. But it does have the advantage of suggesting that elderly and infirm people can receive their own reprogrammed skin cells, and while the cell manipulation and bulking up the new cardiomyocyte population will take weeks, the recipients of the new cells may have plenty of time on their hands. They will have been living with heart disease for years—another ten weeks would hardly be critical. Meanwhile, at the Gladstone Institutes in San Francisco, Deepak Srivastava and his colleagues have gone a step further, but only in a mouse model. They have shown that they can reprogram cardiac fibroblasts (which represent about 50 percent of all cells in the heart) directly into cardiomyocytes, again by firing three genes into them.

The alternative view, that the heart can be coaxed into healing itself, is based on two linked extraordinary findings from the field of transplantation. It was noticed that men who had received a woman's heart, so that all the heart cells should have contained the XX female chromosomes, had some heart muscle cells that bore the male Y chromosome. Conversely, women who had received male bone marrow implants were found to have cardiomyocytes also bearing the Y chromosome. Both results suggested that there is a repair route from bone marrow to heart whereby some cell type can be transported in the bloodstream and turn into a cardiomyocyte. The bone marrow contains a number of different stem cell lineages. There are hematopoietic stem cells, which would normally develop into red blood cells, lymphocytes, and macrophages; mesenchymal stem cells, which would normally develop into bone and fat cells; and endothelial progenitor cells, which would normally form the walls of blood vessels. This knowledge has led to a number of recent human medical trials that introduce bone marrow stem cells into the heart in the hope they will kick-start regeneration.

For instance, researchers at University College London, Barts, and the London NHS Trust have initiated the REGENERATE-IHD clinical trial using bone marrow stem cells for patients who had reached the end of the line for medical intervention for their heart disease. Following a major heart attack, the heart tries to compensate for having lost a substantial portion of heart muscle. In its futile attempt to keep up stroke volume and power, it enlarges and the heart walls begin to get thinner. Cardiologists often refer to a rugby ball turning into something like a football. Patients get more and more tired and breathless, and their ankles swell due to fluid retention. The cocktail of drugs is merely prolonging the inevitable. As Anthony Mathur, who runs the trial, puts it: "The popular conception is that the question you ask the doctor on being told you have cancer is 'How long do I have?' whereas the popular misconception of heart disease is that it is not an end-point, so the opening gambit becomes 'What fancy drugs and techniques can you now employ to keep me going?' But many heart failure patients, now, have lower survival rates than many cancer patients; the goalposts have moved."

Duncan Chisholm and Peter Berry became two of the end-stage heart failure patients accepted for the trial. One-third of the patients were controls in that they did not have their bone marrow taken, and the other two-thirds went into the interventionist arm of the trial. They were injected into the stomach wall for five days with a hormone called granulocyte colony-stimulating factor (G-CSF) to stimulate stem cell production in the bone marrow. G-CSF also stimulates the stem cells to move into the bloodstream and circulate around the body, inevitably to the heart. The interventionist group is then split into three: members of the first group only get the G-CSF, to see what the hormone can achieve for the heart on its own; those in the second group have bone marrow aspirated and injected into the coronary artery and thence to the heart; and those in the third group get bone marrow cells injected directly into the wall of the heart. Peter was warned that the aspiration was not likely to be a pleasant experience: "People had told me that to have your bone marrow out is one of the most painful things you can have. So I braced myself! I'm the sort of person who wants to know exactly what they are going to do and can I see the tools they're going to do it with. One of the nurses said, 'This is the one that goes into your bone!' It's a massive needle—it's like a T-key! But I never felt a thing!"

The bone marrow extracts were purified in the pathology lab to remove any fragments of bone caused by the needle puncture, and the red blood cells, leaving behind the white blood cells and several kinds of stem cell. No further attempt was made to narrow down the choice of stem cell. Stem cells are classified by particular protein markers on their cell surface, but Mathur claims that it is dangerous to assume that you have selected a particular type of stem cell that you have identified on the basis of a particular protein marker because these cells change the way they express these markers over time. You may not end up with what you hoped for. Hence the London team elected for stem cell soup. When the stem cell preparation is returned from the lab, they introduce it into the patient's heart via a catheter pushed in through an opening in the femoral artery in the groin. Although they didn't know it at the time, Duncan and Peter were among those trial patients who did have their own bone marrow cells put into their hearts.

"About six or seven weeks after I'd had the stem cells done, everybody was saying, 'How marvelous you look! You've got color in your face, you're not getting out of breath,'" Peter remembers. Duncan's wife, Rusty, also noticed a change. "I soon began to notice a difference" she says. "He used to get up in the morning and make the tea and then come back to bed, but after about two months, he went off to make the tea and didn't come back to bed. And then, two or three weeks later, he wasn't out of breath by the time he had got up the stairs. He could gradually do more and more without having a problem."

According to Mathur, there seems to be a very real positive effect of the bone marrow injections. The trouble is, they don't really know why. One of the classic ways that cardiologists check for improvement in heart function is by measuring ejection fraction, the efficiency with which blood is pumped out of the heart. However, Mathur's patients, by that measure, showed little or no improvement. Neither was it possible to determine some subtle anatomical improvement in their hearts by examining real-time MRI images. In fact, for a lack of positive results by these measures, bone marrow stem cell trials like the one at Barts are widely deemed to have failed.

Yet some of Mathur's patients are enjoying a real improvement in their quality of life, so what is going on? It is either a massive placebo effect, says Mathur, or the stem cell injections are causing some positive metabolic effect in the heart that they have yet to pin down. This could be some chemical paracrine effect whereby stem cells exude a cocktail of chemokine and cytokine molecules that either squeeze more contractibility out of what functional heart tissue remains, or aid the healing process. Or they may wake up resident populations of stem or progenitor cells in the heart to begin to produce new blood vessels and some cardiomyocytes. Mathur is now collaborating with researchers from heart centers in Switzerland and Denmark in a second related trial, REGENERATE-AMI, which started recruiting in 2008. When a heart attack patient arrives at one of the hospitals in the network, a sample is taken from their bone marrow, once their consent has been obtained, and injected into the heart within hours through a coronary artery that has just been reopened with angioplasty.

At the University of Miami, Joshua Hare is also trying to coax diseased hearts back into life. An American-style research budget has allowed him to do all his pre-clinical research with pigs, which are a much better proxy for humans than mice. Stem cell research has always been bugged by the fact that promising results on animal models translate very poorly into humans. Hare's research has homed in on one single cell lineage from the bone marrow—mesenchymal stem cells. Having induced an infarct in their pigs, they typically injected a total of 200 million mesenchymal stem cells directly into the heart muscle in the zone bordering the infarct. Although ejection fraction of the pigs' hearts did improve, suggesting the left ventricle was getting stronger, more interesting was the actual reduction in the size of the infarct and a reduction in the ballooning of the ventricle. It also seemed that the mesenchymal cell injections had acted as a wake-up call for the small populations of cardiomyocyte progenitor cells thought to reside in the heart. They recorded a twenty-fold increase in the transformation of the resident population of progenitor cells into new cardiomyocytes and blood vessels. Early human clinical trials show similar encouraging results. They are comparing the relative effects of the bone marrow cells, as employed in the London trial, and mesenchymal stem cells. With the latter, they have achieved a 50 percent reduction in infarct scar tissue.

By any standards, Barry Brown would stand out as an extraordinary advertisement for Hare's stem cell technology. He has been a fitness instructor all his adult life, first for the US Air Force, then in the Miami school system, and finally with his own fitness company. But, in 2007, when he was only thirty-eight years old, he quite literally started running into trouble. After about twenty minutes on the treadmill, he was completely exhausted and suffering from chest pains. The doctors discovered that three branches of his coronary artery had become blocked and that he had already had a silent heart attack, robbing him of 70 percent of his heart function. They told him he needed a triple coronary bypass urgently, but they also asked him if he would enroll in one of Hare's trials called PROMETHEUS in which mesenchymal stem cells were injected directly into the heart at the time the surgery was being performed. Four years later, after a slow and painstaking recovery process, Barry celebrated by running a half marathon. He now knows that he received the stem cells and was not in the placebo arm of the trial—and he's not surprised. "I could actually feel my heart healing," he says.

One big practical objection to any stem cell technology for the heart is cost. If stem cells withdrawn from each heart patient have to be bulked up in the lab before being injected into their hearts, the costs are prohibitive for what, Barry Brown aside, are still often very modest gains in heart function. Those costs could be reduced dramatically if the process could be industrialized by purifying mesenchymal stem cells from any human source and expanding those samples into a mesenchymal stem cell bank that would be off-the-shelf. But would it be as safe and effective as using a patient's own mesenchymal cells? Hare is running a trial that directly compares autologous mesenchymal cells (those obtained from the patient's own body) with allogeneic stem cells (those from another donor). A small stage 1 trial has just been completed that suggests allogeneic cells are safe and are not rejected by the host immune system. Larger trials will be needed to prove they can perform equivalently to autologous cells, though evidence so far suggests that hearts do get smaller, scars get better, left ventricles get stronger, and quality of life improves with both treatments, even for patients who had been living with terribly damaged hearts for decades. "We've had people in the study who have thirty-year-old infarcts," notes Hare. "The therapy seems to work even though you have a lot of patients in terminal heart failure with a lot of fibrosis and a lot of heart remodeling. Some of these people were seventy or eighty years old, and even then they were responding to the mesenchymal cells."

In an exciting development of this technology, Hare has returned to his pig model to see what happens if he can remove cardiac progenitor cells from the heart, bulk them up in the lab, and inject them back into the site of the infarct together with mesenchymal stem cells. He has found that injecting both types of cell together has the effect of doubling the reduction in infarct size, but it raises the question of exactly what the mesenchymal cells are bringing to the party. Hare thinks that no single thing is responsible. "Ten years ago everybody assumed that we must have a cell that replaced cardiomyocytes," explains Hare. "That was the underlying thesis as to why embryonic stem cells should work and were preferable. I think we now have a much better understanding from the mesenchymal stem cell studies that it's multi-factorial because we see a cell with a relatively low capacity to differentiate, not zero, but low, still being very effective."

Although the evidence that mesenchymal stem cells can naturally turn themselves into cardiac muscle cells is inconclusive, Hare argues that they must at least be having strong positive paracrine effects through the chemicals they secrete, which can stimulate the growth of new blood vessels and cardiomyocytes and reduce fibrous scar tissue. "What I think is going on is the creation of cardiac stem cell niches," says Hare. "It takes a village rather than an individual. It takes many cells working together to re-create new myocardium, and those cells form niches, clusters of tissue that have ongoing regenerative capacity. That's why, when we combine cardiac progenitor cells with mesenchymal cells, we get more repair because we're replacing more elements of the niche."

There is controversy over evidence that appears to lend weight to Hare's interest in stimulating the heart's own cardiac progenitors. It comes from a joint University of Louisville and Harvard Medical School trial and involves the use of a small pool of cardiac progenitor cells, found to reside in the heart, called C-kit+ cells because they can be recognized by the C-kit protein they express on the cell surface. The original research was conducted on mice in Piero Anversa's laboratory at Harvard Medical School, and the technology was transferred to humans in the SCIPIO trial, headed by Roberto Bolli, in Louisville. Mike Jones was typical of the volunteers. He had been living with congestive heart failure for years and became one of twenty patients involved. He recalls walking into his local convenience store one day and spotting a copy of the local newspaper lying in a rack. It bore a headline featuring Bolli's stem cell trial. Four years after treatment, he describes his life as incredible.

The C-kit cells used in the trial were taken from the hearts of patients during routine coronary bypass operations. The cell population was bulked up to 1 million cells and injected back into the heart at the site of the infarct that had resulted from their heart attacks. Results suggested that scarring was reduced and ejection fraction improved by some 12 percent over one year. Unfortunately, the research has become mired in controversy, with Harvard University demanding retraction of Anversa's key report on his findings, and expressions of concern raised over some aspects of the report on the outcome of the SCIPIO trial.

In the UK, Paul Riley, now at the University of Oxford, has chosen to wake up resident progenitor cells in the heart by a different route. His group had been looking at the problem of creating a new blood supply for the heart after it had suffered an infarction. They identified in mice a small population of coronary precursor cells in the outer layer of heart muscle called the epicardium. Using a small protein, thymosin B4, they were able to induce these cells to form new blood vessels and also triggered them to turn into new fibroblasts (the main component of connective tissue) and muscle cells. While they were plating out these myriad cell types produced by the priming process, they also noticed cells that bore some similarities to heart muscle cells, cardiomyocytes. Could it be that the cardiac precursor cells they were using could not only be a source of new coronary arteries but also of heart muscle cells? In mice, when the cells were primed with thymosin B4, it turned on a classical epicardial marker gene called Wt1. This was not enough to begin the differentiation. It then needed actual injury to the heart to kick-start the process. However, once begun, they were able to track these cells as they developed into cardiomyocytes, saw that they migrated to the site of the injury and that they attached to remaining functional cardiomyocytes, becoming fully electrically coupled with them.

What is the origin of the precursor cells he has discovered? Some researchers believe they do not originate in the heart at all—that they are bone marrow stem cells far from home, induced to leave the bone marrow and migrate in the blood to the injured heart, where they either differentiate into heart muscle cells or aid other cells into doing so. Riley disagrees. He noted that these precursor cells are particularly profuse in biopsies of the wall of the right atrium, taken from sick human patients. This immediately alerted him to the meticulous research of Kristy Red-Horse, from Stanford University, on the embryonic origin of the coronary arteries. According to Riley, ever since Leonardo da Vinci sketched the heart and the surrounding blood vessels, it has been assumed that the coronary arteries bud off from the aorta soon after it leaves the ventricle and then ramify over the surface of the heart: in other words, that coronary arteries have an arterial origin. We now know that this is totally wrong, and the finding may have huge clinical implications for coronary bypass surgery.

Red-Horse has conclusively shown that the cells that go on to form the inner lining of the walls of coronary arteries arise not from arterial tissue, but from a huge embryonic vein, the sinus venosus, which returns blood to the embryonic heart. As the cells journey to the surface of the heart, they are prompted at intervals by chemical signals that point them in the right direction and gradually de-differentiate them from the venous state and re-differentiate them into the arterial state. On reaching the heart, some of the cells in effect burrow into the epicardium and enter the myocardium underneath, completing their differentiation into coronary arteries, while those that remain on or near the surface differentiate into coronary veins, thus completing the coronary circulation. Cells migrate to the aorta, where they bud *in* to form the important connection to the oxygen-rich high-pressure circulation.

Red-Horse has suggested that her insights may prove important for heart repair. Much of the failure of coronary bypass is due to the fact that veins are used. Understanding the chemical processes governing this de-differentiation and reprogramming of venous cells might lead to a new way of inducing fresh coronary artery development in situ, or to tissue-engineering techniques that will grow sections of coronary artery to be used as more successful grafts. However, what piqued Riley's interest in Red-Horse's work is that the right atrium, where he found particularly rich enclaves of cardiac precursor cells, is the mature equivalent of the embryonic sinus venosus at the point it makes connection to the developing heart. Could his precursor cell population and Red-Horse's venous cells be one and the same thing? Could they represent a pool of leftover de-differentiated sinus venosus cells that have lain dormant in the heart for decades?

If so, how could this be turned into therapy? To be an effective aid to recovery for heart attack victims, it would have to work fast, and yet Riley has shown that his precursor cells won't differentiate at all until the injury actually occurs. His idea is to identify at-risk patients (high blood pressure, high LDL cholesterol, and a family history of heart disease would be ideal warning markers) and, rather in the way this at-risk population is today placed on statins, give them a regular dose of an FDA-approved priming chemical, similar or better in action to thymosin B4. This could also be administered to patients at risk of a major heart attack because they have already visited a hospital, perhaps to have a stent fitted, or with a history of angina. The precursor cells in their hearts can then sit in a primed state, waiting against the eventuality of a heart attack, when they can spring into action, invade the ischemic tissue, and help the heart recover its function.

The human heart is an extraordinarily robust and dependable organ. If you live to be eighty years old, it will have delivered over 3 billion powerful, synchronized contractions to power your body with blood. But it is also extremely vulnerable, depending for its own blood supply on the ramifying branches of two small-bore coronary arteries. It is a classic example of the many compromises that evolution has had to make in satisfying the demands on biological design that the forces of natural selection have put upon it. But the more we understand about these compromises, the sooner we will be able to turn them into exciting new therapies to mend sick hearts.

Meanwhile the sheer dogged stoicism of our heart attack victims demands admiration. Duncan Chisholm, age ninety in 2014, is back tending his garden and has returned to the golf course with his pals. "I flag a bit playing golf now. Six holes do me! My pals say, 'You shouldn't be carrying all those clubs, Duncan.' But I give them a touch of the Harvey Smith's [the British horse-rider who became infamous for his rude two-fingered gestures] and just carry on!" Meanwhile, Barry Brown is hard at work, back in Miami, trying to turn inner-city teenagers into top athletes.

Peter Berry finally lost his wife and remained at home on his own, grappling with a range of serious illnesses beside his sick heart. He survived three major operations in less than six months, with 20 percent of a heart, lived alone with a wound in his stomach wall, and a stoma, and, when I met him, was still waiting to see a neurologist for dizziness that made him unsteady on his feet. "You think, 'Why should it beat me?' I'm not suicidal but, at the moment, I'm fed up with life because I can't get out," he told me. Behind Peter's house in Waltham Cross, just north of London, is a wonderful nature reserve, complete with cycling paths. He was itching to get back to it. Sadly his body finally gave up, and he died two days before Christmas in 2013. Although his death was not heart related, his physical heart *had* let him down very badly in life. But his metaphorical heart—his spirit—had been as big as the planet.

THREE SCORE YEARS—AND THEN?

HOW EVOLUTION IS BREATHING NEW LIFE INTO MORIBUND DEMENTIA RESEARCH

Jamie Graham is an imposing gentle giant of a man, six feet four inches tall and sixty-seven years old. He wanders absent-mindedly through his Wiltshire farmhouse occasionally whistling softly through his teeth as if the bewildering predicament he finds himself in beggars belief. Or he sits, head sunk in hands, looking for all the world like someone desperately trying to come to terms with some colossal catastrophe.

Before 2003 he was a globe-trotting IT expert who thought nothing of addressing audiences of two hundred professionals without notes; made light of a strenuous workload; and found plenty of time to be a model father, the life and soul of any party, the perfect home handyman, and an accomplished guitarist with a strong voice and an endless supply of songs from Bill Haley, the Everly Brothers, and Elvis Presley through to Johnny Cash, all stored in his head and reeled off to order each in the appropriate accent. His wife, Vicki, has known him ever since he was fourteen years old. It was the voice that finally hooked her: "We met up again, after childhood, at a wedding party—out came the guitar—I loved his voice—and then there were jokes and laughs—and that was it really." He was the epitome of Shelley's blithe spirit. "He was a brilliant father and always backed me. I remember on a family holiday to the States we were driving across the desert in our old wagon and his imagination came to the fore. He would make the children feel there were Indians about to charge—they nearly fell out of the car with terror. Then peals of laughter—always a spirit of adventure."

Then, one morning, the words on his PowerPoint presentation became a blur and his mind went blank mid-sentence. He started turning up for meetings on the wrong day and appeared stressed and exhausted on his return home from work, fretting over minutiae, tense, and short-tempered. Vicki would overhear him swearing to himself in the tool shed because he'd botched some simple job. "He couldn't even hammer a staple into a post. Even hitching up a trailer he'd get in a muddle." His driving became erratic. He started cutting across in front of oncoming cars and failing to take Vicki's instructions to exit roundabouts. His spatial awareness was disappearing fast, and Vicki found herself, in the passenger seat, perpetually applying mental brakes.

In 2003 Jamie, still in his late fifties, lost his job in London, but in desperation persuaded his employers to grant him a three-year temporary position representing them in the United States. They packed up the house, found a tenant, and decamped to Connecticut. But instead of getting stuck into work and making business contacts, Vicki found him whiling away the hours, playing card games on his computer. The doctor thought stress was to blame and put him on antidepressants, but Vicki wasn't convinced and, in an intuitive moment, marched him off to Nashville, his music mecca, before the premature return to the UK she now felt was inevitable. "Something just told me we had to seize the moment and I have a photograph of him playing the guitar on the stage of the Grand Old Opry, and we went to the studio where Elvis did all his recordings and he stood at the mike where Elvis had sung and sat on his piano stool."

Back in the UK, Jamie appeared to be in some perpetual fog. He still played the guitar a little but started forgetting the words to well-worn standards he should have known backward. In 2005 he came close to estranging a daughter whom he adored because the stress of organizing her wedding, which once would have been meat and drink to him, proved too much. "It was very upsetting for her and me because he kept swearing about 'this bloody wedding.' So out of character." It was Jamie's daughter-in-law Roz, a psychiatrist, who finally realized what was going on. She had been walking with him in the yard and asked him what time it was. He looked at his watch but couldn't make any sense of the dial. Within a year he was taken to London, where a SPECT scan revealed the telltale evidence of amyloid protein plaques in his brain. He was diagnosed with Alzheimer's disease. Vicki remembers, "Jamie didn't react for a couple of weeks and then became really depressed and said, 'I'm dying.' I said, 'You're not. That's the one thing you're not doing.' Although I knew his brain was dying . . ."

For a while Jamie perked up. They were still socializing. "If he got stuck in a sentence, he'd say, 'Well, you see, I've got Alzheimer's!' It opened people up—they were suddenly stunned that this was an acceptable discussion." Vicki packed in as much as she could. They went to music venues, they went skiing, and in 2011 Jamie joined a group of old friends in a charity row down the river Thames in aid of Alzheimer's research. At that time he was still poignantly capable of describing what dementia feels like from the inside. "You just feel left out. I've got something in my head and I'd love to be able to tell you what it is—and I am trying to do that—but you can't get at it. Alzheimer's is what they call it . . . Horrid . . . Go away."

After the rowing episode, Jamie started to go downhill very fast. His world began to shrink remorselessly. He couldn't read and he couldn't use his Blackberry or his laptop. He had been a crack shot, but Vicki now had to squirrel his shotguns away with their son for safekeeping and have his firearms certificate revoked. His driver's license was taken away. He had been a member of a group of countryside ramblers, all with some form of mild dementia or cognitive impairment, who, under supervision, would tramp the Wiltshire footpaths. But that had to stop because Jamie would insist on striking out on his own, and because of his strength the group leaders couldn't control him. He used to bike into the local village to collect the newspapers, but abandoned his bicycle after what Vicki suspects was a narrow shave on the road with a passing car or tractor. He can't accompany Vicki to the supermarket because he makes a nuisance of himself filling other people's shopping carts, and tennis was terminated when he started to fly swat at the ball and brandish his racket in frustrated rage at the unfortunate trainer.

"He has hit me—involuntarily—but if you get on the wrong end of that amount of strength, it hurts. So I burst into tears and tell him, 'I'm your wife! It's me. It's Vicks! What are you doing? Why have you hit me?" And he puts his arms round me and says, 'Oh! Dear! Poor Vicks.'" Jamie can still show empathy toward someone he sees is upset, but he no longer realizes that he is the cause of it.

Disconcertingly, Vicki still sees occasional flashes of the old Jamie—tricks of the light. She says it is rather like seeing the moon appear from behind a bank of clouds. One day an old friend came for lunch whom they had not seen for three years. He had suffered a stroke and was still rediscovering how to walk and talk. Jamie was just mooching about, monosyllabic, recognizing nobody. They decided to go for a short walk. "Jamie came back into the kitchen and looked at this chap and suddenly exclaimed, 'Bill! Yes! Bill! Yes!' And Bill looked at him with genuine delight and they hugged. And then Jamie said his surname! We couldn't believe it."

To their wives' astonishment, the two men shambled off around the garden, arms wrapped around each other, one just discovering how to speak again, the other on the verge of losing speech altogether, looking for all the world as if they were deep in conversation. It was uplifting and heartbreaking at the same time. "I was just so happy that he knew who Bill was and it continued throughout the walk and after they'd left. Sentences were coming out rather randomly: 'Not seen . . . Long time . . .' For a while it was as if some logjam had broken . . . but then the clouds came back."

The statistics on Alzheimer's disease are as terrifying as the sight of a once kind, intelligent, and fluent man disintegrating on a daily basis. It is estimated that 35 million people worldwide suffer from Alzheimer's at an annual cost to health services of over $600 billion. In the United States alone, there are over 5 million cases, with a new case being diagnosed every sixty-eight seconds. Health costs in the United States exceed $200 million a year, with an equivalent amount being spent to fund private care. As people are tending to live longer in all Western societies, we face a medical disaster. All major progress reports on Alzheimer's ominously warn that unless effective prevention and treatment can be discovered, the number of cases is set to triple by 2050. For every decade you live beyond fifty years, your risk of developing Alzheimer's doubles.

Evolutionary biologists look at Alzheimer's disease and suspect that there is little light that evolution can shed on it. This is because most Alzheimer's sets in around your late sixties and beyond. It is the sharp end of old age. Even if you suppose that there are a number of genes that predispose to Alzheimer's disease, it is very unlikely that they can be selected out of the population if their effects only show up in senescent old age, beyond an active reproductive life. This is because evolution depends on the differential reproductive success of individuals in a population, which determines the frequency of any gene, or gene variant, in future generations of that population. In that sense, evolution is not "interested" in the repercussions of this on old age. Any later, deleterious effect of genes in causing Alzheimer's disease, for example, may well have been no barrier to the earlier reproductive success of the individual, and those same genes might actually have conferred some positive effect on survival and reproductive fitness of young individuals that balanced their negative effects in old age. There is therefore no way that selection pressure can operate against such putative genes on the basis that they have late-onset negative effects. There is no evolutionary pressure to protect us against "brain rot," as one prominent researcher put it to me.

However, I believe that the value of evolution to medicine is that it provides us with a forensic approach to try to get to the bottom of what is going on. To look and see whether or not ancient evolved mechanisms are involved, even if they have gone awry in the disease process, or to ask searching and fundamental questions, when looking at structure, physiology or metabolism, as to why such-and-such appears to be so. Alzheimer's research has become worryingly bogged down in recent years, and there is still no known cure for Alzheimer's. The only treatments so far available are merely palliative. I want to show how some researchers, either explicitly or implicitly using evolutionary insights, might be able to rejuvenate the field and lead us to a better-rounded theoretical perspective on Alzheimer's disease and fresh potential for a cure. But first we need to see how conventional research on Alzheimer's disease, known in the trade as the amyloid cascade hypothesis, came about and how it has fared.

Alois Alzheimer first discovered the disease in 1901 when he examined a woman named Auguste Deter at a clinic in Frankfurt. Auguste had presented with a number of psychiatric symptoms, including progressive cognitive impairment, hallucinations, and paranoia. She was frequently delirious and suffered from a "frenzied jealousy of her husband." In 1903 Alzheimer moved to the Royal Psychiatric Clinic in Munich, whose director was Emil Kraepelin. Kraepelin became famous for helping to put psychiatry firmly on a biological basis and first identified the constellation of symptoms that distinguished between the two major forms of psychosis, manic depression, now known as bipolar disorder, and dementia praecox, now known as schizophrenia. When Auguste Deter died in 1906, Alzheimer had her brain sent to him for histological examination. In a lecture based on his findings, he describes the presence of numerous fibrils within even apparently normal neurons, and "miliary foci," deposits of a peculiar material in the superior layers of the cortex. We now understand the fibrils to be made from abnormal highly phosphorylated forms of the tau protein. This is an essential building block for structures inside neurons called microtubules, which act as conduits along the nerve axons for transport of vital molecules. And we now know that Alzheimer's miliary foci are deposits, or plaques, of the insoluble form of a protein called beta-amyloid. Phosphorylated tau and amyloid plaques have since become the classic defining symptoms of Alzheimer's disease.

Kraepelin rushed to bestow Alzheimer's name on this new form of dementia, and there are several theories for his haste. Some argue that he was at pains to wrestle psychiatry out of the hands of the psychoanalysts, while others point to competition from several other laboratories that were turning up similar amyloid pathology in other cases of dementia. Alzheimer is reported to have been mildly embarrassed by the claim to fame foist upon him and found himself a year later writing his own name in description of the eponymous disease after examination of the brain of Johann Feigl, a fifty-six-year-old demented man who had previously been admitted to the psychiatric clinic because he continually lost his way, failed to perform simple tasks, lacked motivation to feed himself, and was generally incompetent. Interestingly, although Alzheimer found numerous plaques in Feigl's brain, he saw no evidence of the neurofibrillary tangles that stood out in the examination of Auguste Deter and have since become one of the two main diagnostic planks of Alzheimer's disease.

Alzheimer's disease first affects the entorhinal cortex, on the underside of the brain, and the nearby hippocampus, which is involved in learning and memory. It causes massive death of neurons in both areas. The disease then spreads to other parts of the cerebral cortex, plaques of beta-amyloid become large and widespread, the brain shrinks, and the ventricles—the large cavities in the center of the brain that are filled with cerebrospinal fluid—enlarge.

As Rudolph Tanzi and Lars Bertram, then both at Harvard Medical School, point out in their short history of Alzheimer's disease research in 2005, clinicians up until the 1980s were still distinguishing between general senility and specific neuro-degenerative Alzheimer's disease by identifying, on autopsy, the classic plaques and tangles that Alzheimer had first described. No progress had been made on either causation or progression, and the search for a possible genetic basis for the disease had proved fruitless. Then, in 1981, Leonard Heston, from the University of Minnesota, first reported a high level of dementia in the relatives of 125 subjects who had autopsy-confirmed Alzheimer's disease. This was consistent with genetic transmission. Heston also showed that Down syndrome, which is caused by the presence of an extra copy (trisomy) of chromosome 21, frequently occurred in these families. Sufferers of Down syndrome tend to develop Alzheimer's at a very early age, and their brains bear the characteristic pathology.

In 1984 George Glenner and Caine Wong extracted and identified a peptide they termed beta-amyloid protein, and this finding was closely followed by the discovery on chromosome 21 of the gene that gives rise to beta-amyloid. The gene, which codes for the beta-amyloid precursor protein APP, was identified by several groups including Dmitry Goldgaber, now at Stony Brook School of Medicine, and Tanzi himself. Thus was the "amyloid hypothesis" born, and it has remained the dominant paradigm in Alzheimer's research ever since. Further research revealed a number of mutations in the APP gene, all of which predisposed to early-onset familial Alzheimer's disease, the type that runs in families, and these were shortly followed by the discovery of mutations in two further genes, presenilins 1 and 2. When APP is cleaved by an enzyme called α-secretase it forms non-amyloid products. However, when it is cleaved by β-secretase, and further cut by γ-secretase, beta-amyloid is produced. The presenilins are subcomponents of this final crucial enzyme complex that completes the pathway from APP to beta-amyloid.

The amyloid hypothesis is based on early-onset familial Alzheimer's disease, which is strongly genetically determined. If you have one of the predisposing mutations, you will get the disease. However, this form of Alzheimer's accounts for less than 1 percent of all diagnosed cases. The vast majority of Alzheimer's is known as late-onset disease, and it is sporadic—it doesn't run in families—which means there must be important environmental influences at play. Nevertheless, researchers like Rudolph Tanzi and Dennis Selkoe at Harvard, and John Hardy at University College London, three important scientists in the field, have strenuously argued that whatever the form of Alzheimer's, the initial event and driving force is the accumulation of neurotoxic beta-amyloid in the brain and that the further development of pathological neurofibrillary tangles of phosphorylated tau protein inside neurons is driven by an imbalance between the production and clearance of beta-amyloid. Therefore, the best hopes for a cure for Alzheimer's must revolve around attempts to curb production of beta-amyloid, by interfering with the enzymes that form it, or by finding a means of speeding up the clearance of amyloid from the brain to make sure it does not build up into plaques.

This modus operandi has dictated over a decade of pharmaceutical research aimed at translating the amyloid hypothesis into drug therapy for Alzheimer's disease. Over two hundred medical trials have so far taken place to test the efficacy of drugs designed either to reduce amyloid production or hasten its departure from the brain. Billions of dollars have been spent on them. None of them, to date, has been successful.

Patrick and Edith McGeer, a veteran husband-and-wife Alzheimer's research team, have launched perhaps the most devastating critique of the amyloid enterprise. Both in their eighties, they are still actively doing research at the University of British Columbia. One of the first clinical trials, in 2003, involved a vaccine against beta-amyloid called AN1792. This was a fragment of beta-amyloid to which a chemical adjuvant had been attached that was designed to stimulate the body's immune system to produce its own anti-amyloid antibodies. Although it had proved effective in preliminary mouse experiments, the human trials were abruptly halted when it was found to have caused death, strokes, and encephalitis in 6 percent of patients. Follow-up of a further eighty patients, say the McGeers, showed no evidence that it had either halted disease progression or cleared tau protein from neurons.

Novartis then designed a vaccine, avagacestat, to target the final enzyme in the amyloid pathway, γ-secretase. It didn't work. This was followed by major trials, involving thousands of patients, with further drugs targeted at beta-amyloid—bapineuzumab and solanezumab. There was some evidence for a reduction of tau and phosphorylated tau, but, overall, they proved ineffective at either removing beta-amyloid or improving cognition. A large international trial of another γ-secretase inhibitor, semagacestat, was halted when patients given the drug fared worse than those on placebo, and some suffered adverse side effects like skin cancer and infections. Attention has now shifted to inhibitors that target the other enzyme in the pathway from APP to amyloid, β-secretase, otherwise known as BACE1. The first results, report the McGeers, are already disappointing. Finally, some pharmaceutical companies have developed drugs that are designed to stop beta-amyloid from aggregating in the brain and turning into fibrils and thence plaques. But, according to the McGeers, tramiprosate, which was hoped would inhibit beta-amyloid molecules from clumping together to form fibrils, was ineffective; and scyllo-inositol, a beta-amyloid neutralizer, either killed patients in high doses or failed to show any effect at low doses. Nevertheless, the international pharmaceutical industry is pressing ahead with trials on a new generation of enzyme inhibitors and amyloid antagonists at a further cost of billions of dollars, despite the fact that the amyloid hypothesis has been rocked by this unexpected failure to translate theory into therapy.

Part of the problem, say the McGeers, is that patients taking part in all these trials are in the advanced stages of Alzheimer's disease and already have massive brain pathology. It may be too late to expect any drug approach to work. It would be helpful if the disease could be caught in its very early stages, but this, so far, is proving impossible because beta-amyloid starts building up in the brain at least twenty years before patients complain of cognitive problems and a diagnosis can be made. By the time, as with Jamie Graham, that you start forgetting things, having mood swings, and finding many tasks beyond you, amyloid and tau have built up, synapses beyond count have been destroyed, the potential for therapeutic intervention has plummeted, and the disease is both irreversible and unstoppable.

Several major trials, in progress, are trying to test the amyloid cascade hypothesis to destruction by identifying cohorts of individuals who bear mutations in the genes that code for the enzymes involved in the production of beta-amyloid. The Alzheimer's Prevention Initiative, headed by Dr. Eric Reiman, has identified a cluster of families near Medellín in Colombia who all carry the same mutation in presenilin 1. Symptoms of Alzheimer's disease commonly show up early in these families—from fifty years old. The researchers are treating one arm of the trial with a new monoclonal antibody, crenezumab, given before any symptoms arise, to see if blocking beta-amyloid production early can lead to prevention.

The Dominantly Inherited Alzheimer Network (DIAN) trial has a similar ambition. Here a global cohort of individuals bearing dominant mutations in either presenilin 1, presenilin 2, or APP have been identified, and, again, one arm of the trial is being treated with drugs aimed at blocking enzyme action—hoping again that early prevention can slow Alzheimer's or stop it in its tracks. These trials may help us to find some form of prevention for Alzheimer's or, conversely, if they do not work, will lend weight to the alternative view that is emerging—that beta-amyloid, and the enzymes that produce it, are the wrong medical targets. A further problem is that all this work focuses on early-onset Alzheimer's, which, as we have said, accounts for only 1 percent of sufferers. It may turn out to be irrelevant to late-onset Alzheimer's, which is a different disease. So, is it the case that amyloid-blocking therapy is a failure because it is not being used early enough, or is it a failure because amyloid is not, on its own, the prime causative agent in Alzheimer's disease? Could it even be because it is the effect rather than the true cause?

Problems and inconsistencies dog the classical, simple amyloid hypothesis. First of all, say the McGeers, it is not at all clear that the toxicity of amyloid has been firmly established. Laboratory experiments to demonstrate that it kills neurons typically use concentrations of amyloid that are a million times higher than exist in the brain. Furthermore, about 25 percent of individuals who die in their late eighties with pin-sharp cognition show substantial plaque and tangle pathology on autopsy. Neither does the extent of amyloid pathology correlate at all well with disease progression. A number of mouse models of Alzheimer's disease, genetically engineered to contain the same amyloid overproducing mutations found in familial Alzheimer's disease, show increased amyloid deposition but no loss of neurons and an absence of neurofibrillary tau tangles. Also, amyloid pathology is by no means specific to Alzheimer's. If you suffer brain damage through a car accident, getting punch-drunk in a boxing match, or because of a wound on the battlefield, amyloid levels soar. Parkinson's disease, Pick's disease, and dementia linked to Lewy bodies—aggregates of alpha-synuclein and other proteins inside neurons—are all associated with amyloid deposition, and amyloid will build up with severe HIV infection, encephalitis, and a number of other severe encephalopathies like Lyme disease, "mad cow" prion disease, or those caused by a variety of brain toxins. Most tellingly, despite the presence of amyloid plaques and tau tangles, you will not get Alzheimer's unless there is also substantial inflammation in the brain, which tells us that the immune system is also involved.

Sue Griffin is professor of geriatrics research at the University of Arkansas. She remembers as a young researcher in the 1980s that the prevailing wisdom in immunology at the time was that the brain was an "immune-privileged organ," meaning that the body's immune system and the central nervous system were completely separate. She didn't believe it, she says, and was not in the least surprised when Roger Rosenberg, at the University of Texas Southwestern Medical School, demonstrated that there were large numbers of brain cells called microglia and astrocytes alongside neurons scattered among the plaques in diseased brains. Microglia were thought at the time to provide structural support and nutrition to neurons, but, to Griffin and others, they bore an uncanny resemblance to macrophages, the scavenging cells of the body's innate immune system. These engulf pathogens and secrete pro-inflammatory chemical messengers called cytokines, like interleukin-1, to stimulate the immune response. "I said there and then, those microglia are going to be making IL-1 and it's going to be doing something to another cell, I figured it was going to be astrocytes, and the two of them would be affecting damaged neurons—and trying to help."

Griffin hypothesized that damaged or stressed neurons would induce the microglia to release IL-1, in turn activating astrocytes to release S100, another soluble inflammatory cytokine, as part of the neuron repair mechanism. IL-1 and S100 were indeed seen to be elevated in the brains of Alzheimer's patients. To progress her research further, she used Down syndrome as a model for early-onset Alzheimer's. Because of the extra copy of chromosome 21, on which the APP gene sits, people with Down syndrome produce more APP than non-affected individuals and develop Alzheimer-like symptoms of plaque and tangles by early middle age. But Griffin discovered that S100 and IL-1 were produced in large amounts in the brains of Down syndrome babies many years before plaques formed, suggesting that the cytokines were indeed released by stressed neurons. Sure enough, when they looked at the more dispersed plaques of early Alzheimer's, they found them surrounded by microglia and astrocytes pouring out S100 and IL-1.

Griffin recalls that the neuroscience community, dominated by exponents of the amyloid hypothesis, was extremely hostile to that early work. She would get papers returned to her with "this is nonsense" scrawled across them. The big question in the minds of skeptics was whether there was a real link between neuronal stress, neuronal repair, and later Alzheimer's. The amyloid lobby, says Griffin, were furious when that link was finally made through a chance meeting with Dmitri Goldgaber, one of the scientists who had first isolated the APP gene. "We fell on each other because what I had done made his finding meaningful. What he had done gave me the link between APP and IL-1. Of course the amyloid people still wouldn't accept it." In an essay titled "My Story: The Discovery and Mapping to Chromosome 21 of the Alzheimer's Amyloid Gene," Goldgaber makes it clear that he had already made the link between amyloid and IL-1 in the 1980s. "We found the level of the APP gene expression was increased by IL-1, an inflammation marker, and thus showed that APP over-expression might be a result of environmental influence. In fact, inflammation was shown to be a major feature in Alzheimer's disease with the increased production of IL-1 in the brain."

Griffin also showed there was a relationship between inflammation and tau—the protein that is vital for stabilizing the microtubules running through nerve axons. Using rats, she demonstrated that IL-1 boosted brain production of the phosphorylated form of tau by elevating the enzyme that performed the phosphorylation, p38 MAPK. This, in turn, proved abundant in neurons from the brains of Alzheimer's sufferers that also had high levels of phosphorylated tau. The link was made.

By 2001, say the McGeers, the key involvement of inflammation in Alzheimer's disease was seen in all its glory. "A variety of molecules, defined as key mediators in peripheral immune reactions, were also found to occur in high concentration in the Alzheimer's brain." Not only did this mean, they explain, that Alzheimer's disease has an important neuro-inflammatory component, but it was evidence, astonishing at the time, that resident brain cells were the source of a wide range of immune and inflammatory molecules, including inflammatory cytokines, and their receptors, together with complement. Complement is a system of small proteins that helps or "complements" the ability of antibodies and macrophages to destroy pathogens. In peripheral circulation, it forms the membrane attack complex, or MAC, which destroys foreign bacteria and viruses by disrupting their outer membranes and literally blowing them apart. It is a major part of the innate immune system.

For Griffin, inflammation in the brain precedes the buildup of amyloid and is a response to neurons that are under stress. For the McGeers, inflammation in the brain is a response, in some individuals, to the buildup of amyloid and tau proteins. "The inflammatory response, which is driven by activated microglia, increases over time as the disease progresses. Disease-modifying therapeutic attempts to date have failed and may continue to do so as long as the central role of inflammation is not taken into account," they contend.

So, if a number of studies throughout the 1980s and 1990s were implicating inflammation and the innate immune system as a major culprit in Alzheimer's disease, why was the converse being ignored, that by reducing inflammation in the brain you might be able to ameliorate Alzheimer's? In fact, a number of retrospective studies had produced very convincing evidence for a positive effect for non-steroidal anti-inflammatory drugs (NSAIDS) like aspirin and ibuprofen. As far back as 1994 John Breitner, at Johns Hopkins University, reported that he had studied a small group of identical twins who all had the Alzheimer's predisposing mutations but succumbed to Alzheimer's at different ages. He discovered that those twins where the onset had been delayed for a number of years had been taking NSAIDS for inflammatory illnesses not associated with dementia. He later did a clinical trial of NSAIDS, the ADAPT trial, on an aged population. He had wanted to use aspirin, says Griffin, but the doctors on the ethical panel refused to allow him for fear the patients would "bleed out," and he was pressured into using two of the then recent-to-market NSAIDS, naproxen and celecoxib. Unfortunately these had unwanted side effects, and so the trial was wound up early. However, the interim results revealed that naproxen, if taken for two years before the onset of symptoms, reduced the incidence of Alzheimer's.

Breitner's work was closely followed by the Rotterdam study that compared 74 patients diagnosed with Alzheimer's against 232 age- and sex-matched controls. They extracted any record of previous NSAID use from the general practitioners' records and discovered that the longer the NSAID use, the lower was the risk of Alzheimer's. Taking NSAIDS for six months or more halved that risk. Finally, a huge Veteran's Administration retrospective study in the United States compared 50,000 former service personnel who had contracted Alzheimer's with 200,000 who were not demented. The taking of ibuprofen for five years or more had halved the risk. Griffin is quick to point out that, while this is all strong evidence for the preventative effect of NSAIDS, it can never be a cure—the disease is simply too complex. And, indeed, in the few trials of NSAIDS in patients with advanced-stage Alzheimer's, NSAIDS have not proved effective. They are preventative, not curative. Griffin believes, cynically, that the lack of research interest in the possibilities of NSAID therapy is due to the fact that the drug companies cannot make much money out of over-the-counter drugs like aspirin and ibuprofen. It is a tragedy, she point out, because if NSAIDS could delay, diminish, or even banish Alzheimer's disease in even 10 percent of the population, the savings would run into billions of dollars a year.

The role of inflammation in Alzheimer's disease remained on the sidelines until the rapid-succession publication of the results of three genome-wide association studies (GWAS) into late-onset Alzheimer's disease in 2009, 2010, and 2011. Philippe Amouyel and colleagues in France performed one study; the other two were masterminded by Julie Williams from Cardiff University. GWAS studies, as their name implies, interrogate the whole human genome for genes that make small individual contributions to a disease process. By enrolling thousands of individuals in these studies, it becomes possible to tease out statistically valid subtle levels of association between gene and disease. To the shock of the Alzheimer's research community, the genes that came out of these highly complementary studies as being most strongly associated with late-onset Alzheimer's clustered into two main groups, those overwhelmingly connected with the immune system and those involved in cholesterol metabolism.

The immune system genes included CR1, which codes for complement receptor 1, a protein that regulates the cascade of innate immune system complement proteins; clusterin, which is involved in inflammation and immunity; PICALM, which may be involved in the way innate immune cells like macrophages (microglia in the brain) engulf debris and pathogens; BCL3, which can regulate inflammation; HLA-DRB1, which presents foreign proteins to immune cells; MS4A2, a receptor for antibodies; SERPIN B4, which modulates immune responses; and many genes connected with the major histocompatibility complex of the immune system. Cholesterol metabolism genes included a variant of the apolipoprotein gene, known as APOE-epsilon 4, which is a known risk factor for late-onset Alzheimer's and can increase your risk of developing the disease by up to sixteen times.

The big surprise for supporters of the amyloid hypothesis, says Williams, came not so much from the genes that dominated the GWAS studies, but from what genes did not feature at all: "That was very difficult to explain from the viewpoint of people who were focused on beta-amyloid or tau as major causal contributors to Alzheimer's disease because when you looked at the common genes—the genes that affect late onset—we had no direct evidence for implicating the APP gene or the tau gene MAPT or even the gene for BACE1 or the presenilins. There was nothing—no variation around those genes at all."

The more genes they found, explains Williams, the more genes they had that implicated immunity as a major factor that caused Alzheimer's disease. The GWAS studies, she says, have changed once and for all the way that immunity is viewed. It is not just a passive response to the presence of amyloid and tau in the brain—as the McGeers posit—but plays a primary role in the development of the disease. We need to broaden our view, she notes. For too long we have put amyloid and tau in a narrow spotlight. The genes that proved so important in familial Alzheimer's disease count for nothing in the dominant late-onset form of the disease that the vast majority of us will experience, and this mirrors the failure of amyloid-lowering drugs to make any difference to the outcome of the disease.

"I'm not arguing that everything about the amyloid hypothesis is wrong because I don't know that," says Williams. "But my problem has been that everything we produce is seen through the lens of amyloid and tau, and we need other lenses to look more broadly across this spectrum of things that are going wrong. Remember, people can live quite happily with a brain that has a lot of amyloid in it, so there's more to it than amyloid and tau. Innate immunity and neuro-inflammation is the big area now which we need to focus on."

It is looking increasingly unlikely that amyloid plaques between neurons, and deposits of tau protein tangles within them, are the initiating events in Alzheimer's disease. They may be, explains Griffin, the rubble that appears as the disease process takes its course. Alzheimer's is a disease of neurons, not amyloid. This mistaken identity of the prime offenders is akin to the air-bag hypothesis, she notes, alluding to a well-known mind experiment on the difference between cause and effect in the context of the motor industry, where researchers discover that in all cases where modern cars are involved in accidents the air bag has been deployed, whereas it is safely stowed, and never inflates, during normal motoring. Thus it is concluded that the main cause of motor accidents is air-bag inflation!

There are many small groups of researchers, worldwide, who have been unsatisfied for years with the amyloid hypothesis and have been trying to find out what really kicks off Alzheimer's disease in the first place. Their work has been out of the spotlight for decades, poorly funded, often derided, effectively sidelined. Although none of them would describe themselves primarily as evolutionary biologists, they are asking the sorts of questions, explicitly or implicitly, that any trained evolutionist would ask of Alzheimer's disease. What are the normal roles for the innate immune system in the brain? Why is it active in the brain if the brain is a sterile organ protected by the blood-brain barrier from the pathogens that insult the rest of our bodies? What causes neurons to sustain damage in the first place? What is amyloid doing in the brain? Is it simply some pathological by-product of brain metabolism, or does it have any normal physiological role in brain function? With the amyloid hypothesis currently in crisis and drug therapy for Alzheimer's disease currently going nowhere, this is a good time to wrestle the spotlight out of the hands of the amyloid lobby and train it elsewhere. That may help to put Alzheimer's disease research on a broader, more pleasing basis, which will allow it to progress more rapidly, and more surely, toward solutions for this terrible disease.

Brian Ross, now in his early seventies, used to be a senior engineer in the British Army Air Corps, overseeing equipment maintenance teams and nursing new technology into service. He was heavily involved in the procurement and adaptation of the French Gazelle helicopter for the British armed forces. He retired in 2004 and about the same time began noticeably suffering from mild cognitive impairment. His condition has recently been diagnosed as Alzheimer's disease, and he has been taking part in a medical trial of an anti-inflammatory drug, etanercept, normally used to curb the painful symptoms of rheumatoid arthritis. His wife, Maree, vividly remembers the day it became obvious that something was wrong with Brian's memory. He had been putting up some shelves in the kitchen and discovered he needed some metal brackets. So they jumped into the car and sped off to their local DIY superstore. "We went into the store and Brian turned to me and said, 'Have we been here before?' and I said, 'Yes. Lots of times—we just want the brackets,' and he said, 'What do you mean—brackets?' He just looked totally bewildered and it frightened me. I thought, 'How can we leave the house with Brian one minute knowing what was going on, and then suddenly he didn't?'"

She seized an opportunity soon afterward to confide her worries to her GP, and Brian went to Southampton for cognitive tests and a brain scan, which showed he had early-onset dementia. He has been on aricept—commonly used to arrest cognitive decline in early dementia—ever since. "We said, 'What happens now?' The doctor said, 'Parts of your brain start to shrink, and the void is taken up with liquid.' We didn't ask for any more detailed information after that because we couldn't really understand it and there was nothing we could do about it."

Now, a decade into the disease process, there are a number of things Brian finds increasingly difficult. He can still drive very confidently but remembering how to get to where he is going has become the problem. "If Maree gave me a shopping list and asked me to go to Sainsbury's supermarket I could cope—once I'd found Sainsbury's!" he jokes. The mental map of the journey has gone. He still enjoys watching documentaries on television, but half an hour after they have finished he has lost all memory of them. He used to love reading but has given up in frustration because he can no longer pick up a book from where he last left off. "Brian reads the newspaper and if he finds something interesting he'll read it out to me. Then, a quarter of an hour later he'll say, 'Listen to this!' and he'll read out exactly the same extract as if it was completely new," recounts Maree. Routine jobs around the house are becoming increasingly difficult. Despite his mild protestations to the contrary, Maree notes that something he must have previously done a million times, like changing the fuse in a plug, is now beyond him. "Brian had lost something the other day, and I told him to look in his coat pocket. He said, 'Where's my coat?' and I said, 'It's in the utility room,' and he said, 'Where's that?' And that really upsets him, and he holds his head then and says, 'What's happening to me?'"

Maree finds it most painful that Brian now has great difficulty remembering enjoyable key past events. She fights the tears because the whole point of a life together is that it revolves around shared treasured memories—memories Brian no longer has. But Brian is a fighter and refuses to retreat into housebound isolation. They make a point of going out for lunch or an excursion every day, and Brian has joined two local choirs, where he sings his heart out even though, to his embarrassment, he can never remember the names of his fellow choristers. "I'm trying to fight it all the time, but it makes me very irritable. And it makes me depressed. I say to myself, 'Well, I don't want to go there, or I don't want to do it,' but you have to fight it—you have to make yourself."

The etanercept study that Brian recently volunteered for, and has yet to be evaluated, is the brainchild of Clive Holmes, professor of biological psychiatry at the University of Southampton. The theory behind it is based on the idea that either mild or acute infections, or the presence of a range of debilitating inflammatory conditions like atherosclerosis, diabetes, or rheumatoid arthritis, increase the levels of pro-inflammatory cytokines in circulation around the body that can then communicate with the brain to induce a parallel inflammatory reaction there. That some initial infection could thus prime microglia in the brain such that, in response to a further infectious insult later down the line or chronic low-grade inflammation, they produce an exaggerated pro-inflammatory response that damages neurons and can therefore become one of the prime movers in Alzheimer's disease.

It all started, says Holmes, when he attended a lecture in the early 1990s in the United States given by Patrick McGeer, who spoke about the connections he was drawing between inflammation and Alzheimer's disease. "At the time everyone was concentrating on the buildup of amyloid in the brain; everything was secondary to that. . . . I came away from the lecture thinking, 'Crikey, that's really important!'" Although McGeer described mechanisms that were restricted to inside the brain, Holmes immediately made the leap to the periphery because, as a clinician, he had experienced large numbers of his Alzheimer's patients reporting that infections seemed to make their condition worse. Could peripheral inflammation somehow influence the course of Alzheimer's?

Soon afterward he was rung up out of the blue by Hugh Perry, who had just moved from Oxford to Southampton to take the chair in experimental neuropathology. Perry had begun a series of experiments to investigate "mad cow" prion disease using laboratory mice. Perry was challenging these mice with a chemical called lipopolysaccharide (LPS), which acts as a proxy for peripheral bacterial infection because it is actually present in the outer coating of bacteria. He found that in mice with prion disease, where there was already tangible loss of neurons, the quasi-infection caused an exaggerated further neuronal loss, much greater than the progression of prion disease on its own. Perry wanted to know whether sufferers from Alzheimer's disease got worse after low-grade infection. Clive Holmes confirmed that they did, but a hasty literature search found nothing to back them up.

They decided on a joint experiment with a few patients where they measured cytokine levels in the blood and interviewed them about their history of infections. There appeared to be a link and so they obtained funding for a larger investigation using three hundred patients. "When I was analyzing all the data," says Holmes, "I suddenly realized, of course, that humans aren't like laboratory animals. Elderly people come with a whole range of inflammatory diseases already; they don't come into the experiment pristine but for the challenge of an infection." There were two things going on—persistent low-grade inflammation in his elderly patients, many of whom had some heart or artery disease, or diabetes, and the occasional acute spike of infection against this low-grade background. They discovered that patients who had chronic inflammatory disease were declining with Alzheimer's disease at four times the rate of healthier individuals. But those with inflammatory disease and a palpable history of infection were declining tenfold.

These results confirmed to Holmes and Perry that states of peripheral inflammation and infection, with their raised circulating levels of pro-inflammatory cytokines, were somehow able to communicate themselves to the brain. However, at the time it was completely heretical to suggest that the brain responded to anything other than neurotransmitters like dopamine and serotonin or that the immune system could "speak" directly to the brain and affect brain chemistry and behavior. It was completely outside the pale of mainstream neuroscience. "Psychoneuroimmunology, as it was then known, was marginalized science," quips Perry, "because people felt that if you were a neuroimmunologist it just meant you were a crap neuroscientist or a crap immunologist! So if you now add 'psycho' to that, it meant that you didn't understand anything!"

However, Perry happened to meet the chief European exponent of this "crap science," Robert Dantzer, at a scientific meeting in France. Dantzer told him about a thirty-year-old evolutionary theory called sickness behavior, which sought to explain the relationship between fever and illness and the almost hibernatory behavior of animals when in the throes of recovery from infections or poisoning. The author of sickness behavior theory, in the early 1980s, was a veterinary scientist at the University of California, Davis, Benjamin Hart. "The most commonly recognized behavioral patterns of animals and people at the onset of febrile infectious diseases are lethargy, depression, anorexia, and reduction in grooming," said Hart. "This behavior of sick animals and people is not a maladaptive response or the effect of debilitation, but rather an organized, evolved behavioral strategy to facilitate the role of fever in combating viral and bacterial infections. The sick individual is viewed as being at a life-or-death juncture, and its behavior is an all-out effort to overcome the disease."

Hart cited research in the 1970s by Matthew Kluger, who had argued that because pathogens generally thrive at a lower temperature than that of their hosts, an ancient adaptation had arisen throughout the animal kingdom to raise body temperature through fever to fight the infection by making conditions for the pathogens inside the body as hostile as possible. At the same time, this adaptation would dramatically reduce activity through sleepiness, depression, loss of appetite, and reduction of water intake, in order to divert the body's valuable resources into stoking the febrile fire. Kluger had deliberately infected rabbits with *Pasteurella multocida*, a pathogen that causes lung disease. He found that rabbits were much more likely to die when given an antipyretic, which reduced their body temperature, than the control subjects allowed to develop fever. The same principle had been used in humans as early as the 1930s by Julius Wagner-Jauregg, who was eventually awarded the Nobel Prize for it. In the days before antibiotics, he found that he could cure neurosyphilis by giving patients malaria, which produced a very high fever. The malarial organism chosen was particularly susceptible to quinine, which was given after the syphilis had subsided. This fever therapy was extended to gonorrhea and remained the only successful treatment for these sexually transmitted diseases until the arrival of penicillin.

Hart specifically identifies three pro-inflammatory cytokines: Interleukin-1 (IL-1), tumor necrosis factor-alpha (TNF-α), and IL-6. These simultaneously cause fever while stimulating immune cells like macrophages to converge on the disease-causing pathogen. These "endogenous pyrogens," as

Hart called them, reset the body's thermostat so that the animal or human feels cold at clement temperatures, inducing it to shunt blood from the periphery to core organs, erect body hair, curl up, and seek out a warmer environment by taking to either a burrow or a bed. He presciently pointed in the direction of microglia in the brain by saying that this acute-phase response—involving fever induction, appetite loss, and increased sleepiness—might be regulated by the central nervous system specifically by neural elements in the brain that contain IL-1.

Dantzer modernized Hart's theory by explaining precisely how signals of infection in the periphery could communicate with the brain to induce sleepiness, social withdrawal, loss of appetite, fatigue, and aching joints. "Cytokines," he said, "convey to the brain that an infection has occurred in the periphery, and this action of cytokines can occur via the traditional endocrine route via the blood or by direct neural transmission via the afferent vagus nerve." In a nod forward to Clive Holmes and Hugh Perry, he states: "Viewing sickness behavior as an adaptive response of the host to infectious microorganisms creates a new and important question: What happens when the acute sickness response is no longer adaptive either because it is out of proportion with the set of causal factors that were the trigger for it or because the sickness response is prolonged? This condition actually occurs during a variety of chronic inflammatory diseases." He specifically cites chronic depression and Alzheimer's disease as two likely repercussions of sickness behavior becoming maladaptive.

Sickness behavior has given Holmes and Perry a solid evolutionary context for their hypothesis that infection and inflammation in the body can raise cytokine levels, initiate inflammation in the brain, damage neurons, and lead to Alzheimer's disease. They believe it is a gradual, chronic disease, accentuated by occasional spikes of signals of infection, and, because nothing dramatic happens in the short term, it is very difficult to pick up reliable signals that something is going wrong. It struck Holmes that a lot of the symptoms that attend dementia also describe sickness behavior—negative apathy, depression, and social withdrawal—and these symptoms turned out to be more prominent in those patients who had chronic low-grade cytokine levels. He is now following patients with mild cognitive impairment, half of which typically go on to develop Alzheimer's, half of which don't, over a five-year period. He is looking to see whether chronic stress, either through adverse life events like the loss of a child or spouse, long-term unemployment, physical illness, severe infection, or chronic pain distinguishes between benign and severe outcomes. His patients are being asked to keep meticulous diaries of life events, and these will be correlated with regular measurements of blood cortisol, inflammatory cytokines, and cognitive decline.

The etanercept study, in which Brian Ross participated, is an extension of this thinking. Etanercept alleviates arthritis because it is a potent antagonist of one of the main pro-inflammatory cytokines—TNF-α. The hope is that by dramatically reducing the signals of inflammation in peripheral circulation, they will reduce those signals in the brain. Holmes is buoyed up by two very recent studies, which suggest that etanercept protects you from developing Alzheimer's if it can be given before the disease has taken hold.

What is it about our reaching three score years that increases our chances of developing Alzheimer's? Holmes thinks it may have something to do with sex hormones. As testosterone levels in men and estradiol levels in women start to drop, inflammation increases. The chronic, low-grade inflammatory responses this sets up in the brain, he explains, are not big enough to stimulate normal negative feedbacks, and so the process just grumbles away all the time. This may give rise to reactive oxygen molecules that damage neurons, causing them to die, and stimulate the production of beta-amyloid. So where, in Holmes's model, do amyloid plaques and neurofibrillary tangles start to arrive? "They start now, I'm afraid," says Holmes, waving toward his brain and mine. "In our brains they are already there. And it may be as a response to chronic low-grade inflammation that beta-amyloid builds up gradually over time. Maybe it then starts to prime microglial cells so that they become very responsive to secondary inflammatory triggers. And round we go."

Perry agrees. It is the switch from an adaptive, evolved, homeostatic process of sickness behavior to a maladaptive process in a diseased brain, when microglia escape from the tight regulation of negative feedback, that the problems start. "If biology has gone to such trouble to regulate these innate immune cells in the brain," he points out, "there has to be a good reason behind it, and it is because, if they are not tightly regulated, they are potentially dangerous and damaging." Microglia, he explains, are not replaced from outside the brain but turn over locally. It is possible they therefore retain a sort of memory of all previous insults so that different exposures in different people with different responses to those insults will determine susceptibility to Alzheimer's. This idea that microglia become "primed" stems from in vitro research, which showed that if you expose a macrophage to a cytokine, wash that stimulus away, and then reapply it, you get an exaggerated response.

Robust support for Holmes and Perry's model for Alzheimer's disease comes from research on mice by Irene Knuesel and Dimitrije Krstic, formerly of the University of Basel in Switzerland. Their work, if proved to extrapolate to humans, dramatically suggests the foundations for Alzheimer's disease can be laid extremely early. Knuesel used a chemical called Poly I:C, which mimics viral infections, and injected it into pregnant mice during late gestation. It caused chronic raised inflammatory cytokine levels in the brains of the fetuses, a reduction in the growth and development of neurons as they matured, and caused a significant increase in APP and some of its breakdown products in the hippocampus (the part of the brain responsible for learning and memory and heavily affected in Alzheimer's) of the offspring as they approached adulthood. They also recognized activated microglia in the hippocampus and an increase in phosphorylated tau protein within hippocampal neurons, which, they suggest, causes nerve synapses—the junctions between nerve cells—to malfunction. Not surprisingly, the mice showed poor performance on Y-maze tests of spatial recognition memory. They concluded that this prenatal immune challenge "is sufficient to trigger a series of pathological events that lead to a slow but gradual increase in amyloid production, tau hyper-phosphorylation, and cognitive impairments, potentially representing a state of increased vulnerability of the brain to Alzheimer's disease."

Things became even more interesting when they gave a second dose of Poly I:C, during adulthood, to mice that had previously been challenged in the womb. This mimicked an adult systemic infection. They discovered widespread changes in both size and morphology of microglia, especially in the hippocampus, which suggested they had become primed. They also saw an abnormal migration and accumulation of astrocytes, the other type of brain immune cell, around damaged neurons. They also found pronounced amyloid plaques in the double immune-challenged mice, compared to controls, that were most prominent in the anterior piriform and entorhinal cortices, which are among the first affected areas in human Alzheimer's disease. They had produced a comprehensive mouse model of Alzheimer's where the sequence of events ran as follows: infection → inflammation in the brain → exaggerated response of microglia and astrocytes → damaged neurons → accumulation of phosphorylated tau inside neurons → accumulation of amyloid plaques between neurons. The classical amyloid hypothesis has always maintained the opposite—that mutations in the amyloid pathway genes give rise to increased metabolism of APP, beta-amyloid, phosphorylated tau, and neuro-inflammation. Knuesel's animal model of late-onset Alzheimer's disease has completely reversed the arrow of causation. It is inflammation that is the prime mover and gives rise to degenerating neurons, an increase in beta-amyloid and phosphorylated tau.

If there is a cautionary motto in Alzheimer's disease research, it could well be "You are only as wise as the last gene you have discovered"! That might well apply to Rudolph Tanzi. His lab was among those that discovered the crucial APP gene mutation that underpins familial Alzheimer's disease in 1986, and he remains a staunch defender of the amyloid hypothesis. However, over the last few years, he has been broadening the status quo to take on board the role of inflammation. As we have seen, the behavior of the immune microglial cells seems key to disease progression. Two genes, CD33 and TREM2, have recently been discovered that affect whether or not microglia perform a benign, protective role or an aggressive role that can lead to Alzheimer's pathology. Tanzi's lab was the first to show that CD33 has a function in the brain, and that the levels of CD33 protein, and numbers of CD33-producing microglia, are increased when Alzheimer's develops. Microglia, as well as producing cytokines as part of the inflammatory response, are also the innate immune system's scavenging cells in the brain. They will gobble up cell debris, damaged and dying neurons, and beta-amyloid by a process called phagocytosis. In this way, healthy microglia help to keep amyloid plaques at bay. In 2013 Tanzi showed that CD33 is both required and sufficient, on its own, to inhibit microglia from gobbling up and clearing one of the most toxic forms of beta-amyloid. They also found the converse to be true—that mouse microglia lacking CD33 increased their ingestion of amyloid and that a mutation in CD33 that reduced the amount manufactured protected against Alzheimer's disease.

Rita Guerreiro, in John Hardy's laboratory at University College London, discovered TREM2, the second gene connected with microglial function. It has the opposite effect to CD33. Expression of TREM2 stimulates microglia to vacuum up cell debris and beta-amyloid, and, at the same time, it stops microglia from overreacting to inflammatory signals, lowers their production of pro-inflammatory cytokines, and steers them toward benign, anti-inflammatory, effective housekeeping behavior. "TREM2 and CD33 are like yin and yang signals," explains Tanzi. "If CD33 is up and/or TREM2 is down, the microglial cells stop clearing the amyloid; they round up into this inflammatory state, start to produce free radicals and cytokines, and instead of being beneficial housekeepers, they start becoming militaristic soldiers. So, at the same time they become neurotoxic, they also stop being phagocytic."

The key to protection against Alzheimer's disease, argues Tanzi, is in keeping microglial cells in a benign state where they are less inflammatory and more effective at removing amyloid from the brain. And, as CD33 and TREM2 have shown us, ultimately whether or not you succumb to Alzheimer's may depend on what innate immunity gene variants you have. Even riddled with plaques and tangles, if a brain stays calm, you will not go down the slippery slope to dementia. But if the brain's immune system overreacts—you will.

The evolutionary story behind Alzheimer's disease has recently taken another twist. It turns out that beta-amyloid is not simply some aberrant byproduct of the enzymatic breakdown of APP; it is actually an active member of the brain's innate immune system. Tanzi remembers sitting in his office a few years ago musing over the growing list of innate immunity genes appearing in the Alzheimer's literature. It was Friday night—beer night—in the Harvard lab, and so he walked next door to visit one of his junior colleagues, Rob Moir, dangling a couple of cold Coronas in his hand. "I said, 'I can't believe all these innate immunity genes popping up on the screen.' And he says, 'Yeah? Well, check this out.'" Moir had been looking at a number of ancient molecules that are effective against bacteria, some viruses, fungi, and protozoans. They are collectively known as anti-microbial proteins and exist throughout the animal kingdom. Humans possess only one of them, LL-37, and he had composed four pages of an Excel spreadsheet on the similarities between beta-amyloid and LL-37, which, like amyloid, can condense into small clumps of molecules called oligomers and insoluble polymer fibrils.

Because LL-37 is a powerful anti-microbial, they wondered if beta-amyloid behaved the same way, and so Stephanie Soscia and Moir tested amyloid against a range of common pathogens including *Candida albicans*, a fungus that causes mouth, nail, and genital infections; *E. coli*, *Listeria*, and *Enterococcus*, all bacterial pathogens; and several species of *Streptococcus*, including *S. pneumoniae*, the leading cause of bacterial meningitis. Beta-amyloid proved effective against all of them, sometimes proving even more lethal than LL-37. They published one paper in the journal *PLoS 1* in 2010 that caused a ripple of interest in the research community but, because it was apparently not followed up, has now largely been forgotten. However, Moir has been plowing on with his research, though he has yet to publish further. "Rob is very careful but also very slow to publish!" says Tanzi. "And one of my jobs is to kick him in the ass and say, 'Hey, man, you've got to get these papers out!' So we now have three papers' worth of data, which I want him to submit in one big paper."

Rob's first experiment involved human neuroglioma cells in culture. Some of them were genetically engineered to express beta-amyloid; some were normal "wild-type" cells. He then introduced a yeast infection. The amyloid-producing cells were completely protected. Using a scanning electron microscope, he found that the amyloid had condensed into balls of fibrils that he calls nano-nets, to trap the yeast cells. It then interacts with active metals like copper to release a cloud of toxic free radicals that attack the yeast cells, literally punching holes in their membranes, and thus destroying them.

Their next infection model used a nematode worm, *Caenorhabditis elegans*, and attacked it with yeast. "The result, under the electron microscope, is quite brutal," notes Tanzi. "The yeast goes into the *C. elegans*, and it's like the movie *Alien*—they pop out of the *C. elegans*' guts and kill it from within." However, again, if the worms were engineered to make amyloid, they were unharmed.

Finally, they selected a strain of mice genetically engineered to express the APP and presenilin gene mutations common to familial Alzheimer's disease, which, as we know, will lead to a buildup of amyloid. They then injected the *Salmonella* bacterium into the hippocampus of the mouse brains. Wild-type mice, which did not contain the mutations, died within a few days, but the "Alzheimer's" mice lived twice as long. Evolution has employed beta-amyloid as a very potent anti-microbial agent in human brains, and the downside is that the very toxic nature of beta-amyloid, capable of forming mesh-like clumps to imprison microorganisms before putting them to death with free radicals, means that it has also produced a mechanism with potentially very damaging side effects. If it goes wrong, there is nothing to stop amyloid from turning against neurons.

I hope that you might already have put two and two together to ask, "Why would evolution design a very nasty microbe-fighting device, with potentially lethal fallout in the neuronal damage that can lead to Alzheimer's disease, if the brain is a sterile organ protected from infection by the impregnable blood-brain barrier?" We will get to the answer, but first we need to introduce another example of an evolved mechanism in the human brain that has unwanted side effects that can lay the groundwork for Alzheimer's disease.

In a process beginning before birth and continuing throughout childhood, our brains grow quickly and are filled with a rapidly growing number of neurons, connected together by their synapse junctions to form complex neuronal networks. This leads to overgrowth, a super-abundance of neurons and their interconnections. During adolescence this overgrowth is selectively pruned back to establish precise, efficient, neural circuits, and in the process under-performing synapses are destroyed. There is evidence that this neural plasticity continues into adulthood on the "use it or lose it" principle that underused or relatively inactive neural circuits are nipped off and other neural interconnections strengthened. Evolution has co-opted the innate immune system in the brain to do this synaptic pruning.

We know that a cascade of complement proteins is used throughout the body to tag unwanted cell debris and invading pathogens so that they can be recognized by macrophages and gobbled up. In the brain, the complement cascade is similarly involved in innate immune processes. But it is also heavily involved in the production of new neurons from stem cell progenitors either during development or as a response to injury, the migration of neurons to their appropriate stations in the brain, and the sculpting of synapses. During brain remodeling, it coats synapses that are destined for destruction, which attracts microglia and, later on, astrocytes, to engulf and destroy them. Synapses are initially painted with a complement protein C1q, which interacts with the surface proteins on cells and debris to form C3. This is recognized by the C3 receptors on microglia, causing them to steam in for the kill.

Beth Stevens, at Harvard, and her colleague Ben Barres, now at Stanford, believe this process can reawaken in later age and is responsible for the pathological synapse destruction and thus neuronal loss that occurs decades before the cognitive symptoms of Alzheimer's disease arrive. Most body cells are protected against unwanted attack by complement because they are bristling with strong inhibitors. Neurons lack these inhibitors; they have evolved to be selectively open to complement and microglial attack, otherwise pruning could not occur. They are thus left with an Achilles' heel later on and unable to resist complement in the pathological context.

Barres has recently shown that C1q normally accumulates up to 300-fold in brains as they age, most of it local to synapses, but it requires a second hit to activate it and begin synapse elimination. What might that second hit be? Brain injury, says Barres, is an obvious example. The trauma sends the innate immune system into overdrive, and head trauma may temporarily open the blood-brain barrier, allowing peripheral sources of complement to pour in, along with any microorganisms that have infected the bloodstream. Systemic infection is another likely bet. Barres and his colleagues are extremely intrigued by the work of Holmes and Perry on the rate of cognitive decline in Alzheimer's patients following peripheral infection. This is exactly what they would predict because they believe that peripheral infection primes microglia and astrocytes in the brain, which then start to pour out C3 to drive synapse loss. Barres and his colleagues have mimicked such peripheral or systemic bacterial infections in experimental animals using injections of lipopolysaccharide and shown that C3 production is dramatically increased. They think this complement-driven neurodegeneration can start unbelievably early in life. For years the patient may remain without symptoms of cognitive decline because the brain can compensate by forming new synapses as fast as they are being destroyed. It is only when the complement fire, burning and spreading through the synapses, becomes overwhelming, they argue, that cognitive impairment shows up.

So, if trauma or peripheral infection can provide the trigger for massive complement production in the brain, what tells it to target particular synapses? The most likely candidate is beta-amyloid. Synapses are in a constant state of homeostatic control. If the traffic of nerve impulses across them and through neural networks is unrestrained, it may result in hyperactivity, which we might experience as a fit or seizure. In the opposite direction, learning and memory depend on the long-lasting transmission of nerve impulses across a particular network of neurons. This phenomenon is called long-term potentiation (LTP). We now know that beta-amyloid produced inside neurons is a major part of the negative feedback that impairs LTP. Tanzi believes it possible that if anything disrupts this delicate feedback mechanism, if beta-amyloid accumulates in neurons, for instance, the impaired traffic across that synapse might send out a signal to microglia that it is weaker than the next one and that "they need to eat that synapse."

Tanzi's working hypothesis is that beta-amyloid, in normal physiological concentrations, has a dual role as a protective molecule in the brain. "Let's say you have a head bang—a trauma," he explains. "As part of the acute reaction, beta-amyloid levels go up at the site of injury and two good things come about: One, amyloid shuts down the grid—it's turning off or tuning down network activity. Second, you may have a blood-brain barrier that is compromised and you have pathogens leaking into the brain. The same molecule—beta-amyloid—fights the pathogens. This is why we think beta-amyloid evolved as an acute-phase protein, to serve this dual purpose."

Tanzi believes that, through this evolution of beta-amyloid as an acute-phase protective protein, evolution has been playing with fire. Evolutionists call it antagonistic pleiotropy. What enables you to survive when you are young may contribute to your demise when you are older—you live now and you pay later. Anything that allows beta-amyloid to accumulate, he argues, risks the fire burning out of control. In fact, as long ago as 1992, Joseph Rogers showed that beta-amyloid binds to complement C1q, activates it, and gives rise to the final pathological product of the complement cascade: the membrane attack complex, or MAC. This may attack faulty neurons, fatally damage them, and call in microglia and astrocytes to finish them off.

It is an extraordinary fact that much good research was done twenty or even thirty years ago into the way synapses and neuronal networks function, and how dependent they are on physiological concentrations of APP and the whole family of APP metabolites, including beta-amyloid. That work has since been totally eclipsed by the amyloid hypothesis. For instance, in 1991 R. D. Terry showed that dramatic synapse loss correlated with the severity of Alzheimer's disease. His group was the first to develop techniques to count cortical synapses in autopsy tissue from normally aged and Alzheimic brains, and he found only very weak correlations between performance on psychometric tests of intelligence and the density of Alzheimer plaques and tangles—but much stronger correlations with synapse density. He was awarded the Potamkin Prize for this path-finding research in 1988. Others have shown that APP is necessary to form synapses in the first place and to regulate their strength, and that, under normal conditions, the majority of APP does not go down the amyloid route at all but is cleaved by a different enzyme, α-secretase, to form a family of non-amyloid products that are vital for protecting neurons, controlling their excitability, and governing the growth and branching of nerve axons.

Beta-amyloid is very complex multifaceted stuff. It is found throughout the body as well as the brain and is, according to Caleb "Tuck" Finch, professor of gerontology at the University of Southern California, part of an ancient evolutionary mechanism for wound healing and inflammation that long preceded the arrival of the adaptive immune system with its immune memory and the production and targeting of specific T lymphocytes. Inflammation, he explains, goes right back to the beginning and can be seen across the invertebrates, including insects, crustaceans, annelid worms, and echinoderms. Why only humans seem to develop Alzheimer symptoms of neuro-degeneration, including plaques, tangles, and neuron loss remains a mystery, he notes. Nor is there an adequate explanation for why the brain has no adaptive immune system but has, instead, drawn on ancient components of the innate immune system, like beta-amyloid and complement, and pressed them into vital new roles in building and protecting the brain, roles that leave the brain cruelly exposed to the unwanted effects of complement and amyloid in later life. It is a quick and dirty evolutionary fix with potentially devastating consequences.

In the end, as Daniela Puzzo, of the University of Catania, has commented, beta-amyloid is a Jekyll-and-Hyde molecule depending on its form and its concentration. At low concentrations, it has positive effects and may even be vital for the long-term potentiation on which learning and memory depend; at higher concentrations, it may protect synapses from self-inflicted overexcitement damage by depressing LTP; but at higher concentrations still, it can depress traffic across synapses too much, weaken them, and threaten their survival. All this work on the normal physiological role of beta-amyloid urges caution in applying any drug regime designed to reduce its production—instead of protecting against Alzheimer's disease, it may actually do more harm than good.

The view of Alzheimer's disease we have been arriving at starts with something dysfunctional happening at the synapse. This spreads throughout vulnerable areas of the brain, resulting in massive synaptic and neuronal loss, cognitive decline, eventual buildup of amyloid plaques between neurons and phosphorylated tau protein tangles within them, a diagnosis of Alzheimer's disease, growing confusion, disability, and death. We have largely turned accepted wisdom on its head. The various theories to explain the original loss of synapses have invoked head injury, possibly with the breaching of the blood-brain barrier that allows pathogens into the brain; normal aging, which also makes the blood-brain barrier less effective; or peripheral signals of infection, which, when communicated to the brain, ring alarm bells in the microglia along the lines of "Watch out! Something evil this way comes!" But we have been skirting around one obvious conclusion as to what might be the original, precipitating factor that kicks off Alzheimer's disease and damages neurons. This is the idea that Alzheimer's disease actually begins with long-term infection of the brain with bacteria, viruses, or other pathogens. In their paper on the role of amyloid as an anti-microbial protein, Soscia, Moir, and Tanzi mention studies claiming to show infection of the central nervous system with *Chlamydia pneumoniae*, *Borrelia spirochetes*, the bacterium *Helicobacter pylori*, and a number of viruses. Nevertheless, this "pathogen hypothesis" is one of the most contentious of all the conjectures surrounding Alzheimer's disease, and its proponents have been branded heretics. But I think that if you argue from basic biological principles, it is difficult to avoid the conclusion that the "pathogen hypothesis" merits greater attention, and this is why I want to end this chapter with a brief survey of the research that points to it.

Brian Balin is professor of pathology, microbiology, immunology, and forensic medicine at the Philadelphia College of Osteopathic Medicine. He has become exasperated with the politics of Alzheimer's research. "You've got some very prominent labs that are as thick as thieves with pharmaceutical companies," he says. "The companies are using the heads of those labs as their advisory groups, and they are telling them, 'This is where the answer is, this is what you have to put your money behind for the research. Fund us for work on amyloid, and you'll be able to use these targets we can give you for drug development.' And all the little fringe groups—like us—are saying we are studying the problem based on biology, and what makes sense to us is not the amyloid story. It never has been."

Balin points out that a proof-in-the-pudding example for the pathogen hypothesis has been staring the community in the face for decades. It is widely accepted that the HIV virus can enter the brain by piggybacking inside white blood cells in the bloodstream, which then manage to squeeze through the blood-brain barrier. Inside the brain, the virus rapidly infects microglia and astrocytes, which start to spew out pro-inflammatory cytokines and other toxins that destroy neurons across the cortex and particularly in the hippocampus. HIV dementia is an acknowledged result of these infections and was widespread before the use of effective anti-viral medicines for HIV-AIDS sufferers.

Balin's team have been investigating *Chlamydia pneumoniae*, a bacterium that lives inside cells and is a major cause of pneumonia. Balin has longed mused over the fact that one of the earliest faculties to be damaged in early Alzheimer's is one's sense of smell. The lining of our upper noses, where all our olfactory receptors are located, is the perfect point of entry into the nervous system for a virus or bacterium, explains Balin, because dead and dying cells are being replaced with new cells in rapid turnover every ninety days.

On top of that, you have a constant bombardment of the epithelium with airborne toxins and microorganisms, along with the inflammation of sinusitis. So the epithelium is leaky. Balin believes that, because of the relatively small size of the infectious part of the microorganism, *Chlamydia* can enter nerves in the nose and travel into the olfactory bulb of the brain, and thence to the entorhinal cortex and other parts of the brain affected in Alzheimer's disease. It is for *Chlamydia*, he notes, the royal road into the brain. He has shown the presence of the organism in those nerve pathways in human and animal models following inoculation into the nose.

In vitro research has also shown a second route, identical to the path chosen by HIV, where *Chlamydia* can smuggle itself into the brain. It can travel, Trojan horse style, inside white blood cells, where the body's immune system cannot get at it. If *Chlamydia* can first get to the lungs, it can be picked up by white blood cells in lung blood capillaries. Once in the brain, infection spreads to microglia, neurons, and astrocytes, and they all send out "Alert! Invaders have arrived!" cytokine messengers, which recruit further, possibly infected, white blood cells into the brain—a second hit. The resulting infection and inflammation can then stimulate amyloid and give rise to the visual pathology of Alzheimer's disease. This makes amyloid the effect not the cause. As infected neurons die, *Chlamydia* leaks out into the extracellular brain tissue, where Balin's group has identified it using electron microscopy, together with chemical proxies for *Chlamydia*, including lipopolysaccharide molecules from the bacterial coat, and specific antibodies. The bacterium can then go on to infect further microglia and neurons.

Balin points also to the dental research of Angela Kamer and colleagues, from New York University. In the mouth, explains Kamer, plaque-induced gingivitis, which often occurs in young people, can be treated, but the more severe periodontal diseases are irreversible, highly inflammatory, and result in tooth loss. They spread from the gum line into deeper connective tissue, which soon develops deep ulcerated pockets containing a seething mass of bacteria, leukocytes, macrophages, T and B cells, and a multitude of cytokines and chemokines. At least half of Americans over the age of fifty-five, she says, have chronic periodontitis, which can be a constant source of circulating inflammatory cytokines capable of communicating with the brain. And the organisms themselves—like *Actinobacillus*, *Tannerella*, *Porphyromonas*, and *Treponema*—can also get into the bloodstream. Although there is some disagreement between studies done so far, the spirochete *Treponema denticola* has been found in the trigeminal nerve ganglion, situated close to the brain. Another study found *Treponema* in the majority of Alzheimer's disease brains, but in very few controls. Kamer herself has found that antibodies to periodontal bacteria are indeed elevated in Alzheimer's patients versus controls. Kamer cautions that her studies, so far, have been of small size and are not yet fully capable of distinguishing between cause and effect, and pleads for larger longitudinal studies to affirm the link.

To be a believable candidate for an infectious cause for Alzheimer's, any microorganism must be shown to be a widespread rather than occasional pathogen, and Balin's group have shown by serological sampling of older human populations that between 70 and 90 percent are positive for *Chlamydia pneumoniae*, and that in a population of cognitively impaired people over sixty-five, the majority of white blood cells are infected. The big question is—has the microbe entered their brains?

Ruth Itzhaki, together with Matthew Wozniak and colleagues at the University of Manchester, has been investigating links between Alzheimer's disease and the herpes simplex virus, HSV-1, the latent virus responsible for painful, recurring attacks of cold sores. She was led to the possibility of a connection by noting that the regions of the brain that are affected in Alzheimer's disease are the same as those affected in the mercifully rare herpes simplex encephalitis, which causes memory loss and cognitive impairment. Following a viral outbreak resulting in a painful cold sore on the lips, she explains, the virus retreats along a branch of the trigeminal nerve to the trigeminal ganglion, which lies inside the skull in a cavity in the dura mater, the tough, fibrous covering of the brain. Here it lies dormant until a period of stress, another systemic infection, or even old age weakens the immune system, which is corralling it inside the ganglion, and another opportunistic outbreak occurs. The trigeminal ganglion has connections into the brain stem and the nearby temporal cortex.

In Oregon, Melvyn Ball and colleagues have been researching the role of HSV-1 since 1982. A recent report, he notes, describes Alzheimer's disease as "spreading like a viral infection from brain cell to brain cell." The authors were using a figure of speech to describe the "infection" of Alzheimer's, which they believed was caused by the transmission of phosphorylated tau from neuron to neuron. But Ball scoffs at that idea. Phosphorylated tau tangles inside cells, he explains, resemble forkfuls of cooked spaghetti. It is hard to imagine any highway between neurons that could physically transmit them. You need a bullet, says Ball, and he believes that bullet is a virus—

HSV-1. Because the trigeminal ganglion lies so very close to the limbic centers of the brain, if the virus moves out of the ganglion toward the center, it will find itself in the brain, rather than finding itself on the lips by moving to the periphery. Virtually 90 percent of North Americans, Ball points out, harbor HSV-1 in their trigeminal ganglia, and he has identified HSV-1 in 67 out of 70 postmortem brains of confirmed Alzheimer's disease victims. One of his researchers has identified a gradual spread of lesions best interpreted as the neuron-to-neuron spread of infection.

Ball is a skilled micropathologist. It is generally accepted, he says, that a viral trigger causes the neurofibrillary tangles seen in Parkinson's disease patients, and that in a rare neurodegenerative disease, panencephalitis, the measles virus genome can be found in tangle-laden nerve cells. Ball took a series of photomicrographs from the brain of an eighty-seven-year-old man who had died after fourteen years of progressive dementia and showed a microglia shrub or nodule surrounding and engulfing a dying brain-stem neuron containing a typical tangle. It is accepted that this microglial siege is typical of cerebral viral infection. Both Itzhaki and Ball are backed up by veterinary research by Maxim Cheeran at the University of Minnesota, who has tracked HSV infection into the brain from the trigeminal ganglion in a mouse model of herpes simplex encephalitis. He noticed activated microglia congregating around infected areas, giving rise to "long-term smoldering inflammation." Maze tests on the infected mice revealed loss of spatial memory that is similar, notes Ball, to the spatial disorientation experienced in the early stages of human Alzheimer's disease, as in a man who has trouble finding his way home or locating his car in a parking lot.

Finally, Itzhaki claims to have shown, in vitro, that anti-HSV agents were able to reduce the phosphorylated tau tangles in neurons that she had previously induced to accumulate by deliberately infecting them with HSV-1. She also showed the opposite—that tau accumulation depends on the replication of HSV-1. Meanwhile, Ball and his colleagues have demonstrated that HSV-1 associates with beta-amyloid lesions in the brain and that in vitro infection of glial cells with HSV-1 in culture produces substantial pathology of neurons that can be held in check by either application of soluble beta-amyloid or acyclovir, a proprietary antiviral drug often prescribed for herpes infections.

Much more research needs to be done on the potentially important pathogen hypothesis for Alzheimer's disease, and there is far too much we do not know about the normal physiological roles of APP, amyloid, and complement in fighting pathogens and supporting neurons, especially as they grow old. And we do not really know whether it is infection that is the fire that spreads through neural networks as Alzheimer's disease progresses, or whether it is tau, or whether it is actually synaptic dysfunction itself that infects other synapses and thus neurons. So myopic has the amyloid hypothesis been that the research community has, for instance, made little headway concerning the one gene we *do* know increases your chances of contracting late-onset Alzheimer's disease by up to sixteen times: APOE, apolipoprotein E, exists in three genetic variants, epsilon 2, 3, and 4. It is the epsilon 4 variant that carries the risk, but no one really knows why. Julie Williams, the lead author on the two big, revolutionary GWAS studies on Alzheimer's disease, thinks this research omission is nothing short of criminal. "It is the one thing that distresses me more than anything—how few people are working on APOE. It's ludicrous, actually. Whenever I go to conferences, I look for anybody working on APOE and they are rare. Yet it is an enormously strong association that we desperately need to understand. It's illogical, irrational, and damaging to our understanding of Alzheimer's disease."

There are, in fact, a number of groups working on the association between APOE and Alzheimer's disease, but Williams is correct to point out that the research, to date, is very inconclusive. Rick Caselli, with a number of Alzheimer's research colleagues in Arizona, has recently reviewed the efforts of the research community in trying to discover the exact nature of the link between APOE epsilon 4 (APOE ε4) and Alzheimer's. While much current research indicts APOE in the context of its posited interplay with beta-amyloid, Caselli notes, evidence is emerging for a number of processes involving APOE that are entirely independent of beta-amyloid and may be just as important—or even more so.

For instance, APOE is responsible for the transport of cholesterol and other lipids in the brain that are vital for the development and maintenance of neurons and their synapses. This transport is impaired in experimental animals carrying the APOE ε4 gene variant, which may affect neuronal health. APOE ε4 has also been implicated in the phosphorylation of the tau protein inside neurons, which may help lead to the classical neurofibrils of phosphorylated tau long recognized as a prime pathological marker of Alzheimer's. These neurofibrillary tangles of phosphorylated tau have been shown to affect learning and memory in laboratory mice. Preclinical cognitive decline, especially memory loss, in humans is linked to APOE ε4, as is impairment of neuronal development resulting in thinner entorhinal cortices and reduced hippocampal volumes—the two areas of the brain that first succumb to Alzheimer's disease. APOE ε4 may also damage the function of the mitochondria—the energy powerhouses inside neurons and all cells—and harm the astrocytes, which provide essential nutrient support to neurons. Vascular blood flow inside the brain is impaired in human APOE ε4 carriers, and there is breakdown in the vital blood-brain barrier that normally tightly controls what molecules can traffic into and out of the brain. And APOE ε4 is frequently demonstrated to increase inflammation in the brain, particularly through the stimulation of a variety of pro-inflammatory cytokines, prostaglandin, and primed, toxic microglia.

Caselli points out that, fascinating though these clues to Alzheimer's pathology are, the work is done in a limited number of laboratories and is consequently either inconclusive or urgently needs replicating and extending by wider scientific support. Meanwhile, several researchers, including Itzhaki, have discovered an association between Alzheimer's disease, HSV-1 infection, and the APOE ε4 gene variant, which they believe increases infectivity because it allows the virus more easily to invade neurons and microglia and spread throughout the brain.

Although the pathogen hypothesis is attractively argued, the links between the presence of pathogens in the brain and the onset of disease will have to be thoroughly nailed down before they will be taken seriously by the wider Alzheimer's research community. But Ball is surely correct when he pleads that the cost to mount one major human trial of a common, relatively cheap antiviral agent like acyclovir is tiny given the vast amount of money already spent annually on the amyloid hypothesis. As Ball says, "All investigators in the field of medical bioscience contribute very critical building blocks to the frustratingly slow progress that inevitably predates any final disease breakthrough, and as in other realms of competitive human endeavor, bioscience has its fashionable bandwagons. We now implore federal and other funding agencies to immediately allocate at least half of all new Alzheimer's disease budgets to younger, smaller laboratories burning the midnight oil to find the missing link responsible for the enormously tragic and costly burden of this disease."

Ball's criticisms echo a remarkably candid and devastating commentary on the amyloid hypothesis a few years ago by Rudy Castellani, George Perry, and Mark Smith of Case Western Reserve University in Cleveland and the University of Texas. Perry is the respected editor of one of the most influential histories of Alzheimer's disease research. The three researchers argue:

> The perversion of the scientific method and manipulation of a desperate public afflicted by an expanding, devastating, and incurable disease, characterize Alzheimer's research and treatment in the twenty-first century. Nevertheless, the peer-review processes, the competition for public funds, and the embedded centers of opinion continue, unabated and unabashed, as progress toward effective treatment stagnates. . . . Pathological interpretation of neurodegenerative diseases has, for better or worse, focused on plaques and tangles for no other reason than the fact that they can be visualized, and implicit in the vast majority of studies on Alzheimer's disease is the assumption that these lesions themselves are inherently toxic, and therefore represent disease per se rather than disease response. Remarkably, after nearly a quarter century of lesion primacy, and after targeting of beta-amyloid as an avenue of treatment has repeatedly failed, the amyloid cascade concept is referred to as "once controversial."

"At some point it has to come down to 'Where is your return on investment?' to continue on in that vein," comments Brian Balin. "And it goes also to the return on investment for the population that's getting the disease. If it is wrong, you have to say that and move to something else."

Spare a last thought for Brian Ross, gamely fighting against the dementia that is slowly but surely engulfing him. Or Jamie Graham, reduced from a talented businessman to a bumbling wreck of a man. As his wife, Vicki, says: "For me it's like being a widow. I've known and loved him man and boy for over fifty years, but now I'm living with this husk of a man walking around who looks like Jamie but is not there."

Millions of people like Brian and Jamie are living in the twilight world of Alzheimer's disease without the ghost of a cure in sight. It is a hideously complex disease, but it has become transparently clear that the amyloid hypothesis alone cannot account for it. To their credit, many prominent researchers from the amyloid camp—like Rudy Tanzi, John Hardy, and Dennis Selkoe—have broadened their horizons, as Julie Williams implores them to do, to shine the research spotlight elsewhere. Biology and evolution have led us to explore new ways of injecting life into a moribund area of disease research. They point up the extraordinary way that evolution has invested in potentially dangerous mechanisms for sculpting human brains in early development and protecting young brains from traumatic insult. All this is at the expense of cognition in older people approaching or past active reproductive age.

It makes me wonder why we "oldies" are not all tottering around with Alzheimer's disease; and given that infection is a constant fact of life—together with the potential for heart disease, diabetes, obesity, and other inflammatory ailments afflicting our later years—it may finally come down to what gene variants each of us is endowed with as to whether we are somewhat protected against the inflammation that fires up dementia or if we are left susceptible to it. Even though we now understand much more about the primary roles of inflammation, infection, and the innate immune system, the most intractable problem in Alzheimer's disease research is how all this will translate into preventative medicine. The seeds of Alzheimer's disease are sown much earlier in life than we once thought, and we have at present no biological markers that can tell us something is going awry decades before cognitive problems surface.

Whatever really is at the root of Alzheimer's disease—microorganisms, inflammation, disease of neurons and synapses, and if indeed Alzheimer's is the price many of us are paying for living long enough to suffer the downside of evolutionary fixes that made young human brains such formidable organs, or protected young brains from injury and pathogens once rife in yesterday's much dirtier and dangerous world—we will have to find some way of detecting Alzheimer's disease at outset or face the consequences in a worldwide pandemic of dementia from which there can be no recovery. Whether we beat it or not might depend on moving a heretical band of scientists, arguing from biological first principles about the peculiar way that evolution has worked to blind design the human body, from the wings to center stage of Alzheimer's disease research. We have no time to lose.

ACKNOWLEDGMENTS

Many medical scientists have kindly given me valuable time, help, and access to the details of their research. I am deeply indebted to all of them, and the list is too long to mention all by name. However, I must reserve special thanks to a number of researchers who have been especially generous with their time and material aid. Andrew Wragg and Anthony Mathur and terminal heart failure patients at the London Chest Hospital; John O'Dowd and Jeremy Fairbank for advice on orthopedic surgery and access to patients, and Mike Adams for advice on biomechanics; Graham Rook for advice on the immune system and the hygiene hypothesis; Ronald Kröger for alternative views on eye design, and Robin Ali for help with stem cell regeneration for the eye and kind permission to use his title "DIY Eye" for my chapter; Jennifer Ackerman for kind permission to use the title "The Downside of Upright"; David Haig for detailed correspondence on parent-offspring conflict theory, and Jan Brosens for showing me how he explicitly frames his work around Haig's theories, and Ian Sargent and Chris Redman for the background to preeclampsia; Colleen Farmer for advice on the evolution of the vertebrate heart; Colin Watts for access to a brain tumor operation; Mel Greaves, Carlo Maley, Brian Reid, Ian Tomlinson, and Trevor Graham for examples of the application of evolution to cancer research; Sue Griffin, Rudy Tanzi and Rob Moir, Julie Williams, Clive Holmes and Hugh Perry, Irene Knuesel, Ben

Barres, Ruth Itzhaki, and Brian Balin for helpful discussions on Alzheimer's disease. Several of these scientists have generously read and critically reviewed individual chapters, and I particularly wish to thank Randolph Nesse for helpful comments covering all chapters. Their help was invaluable, and, needless to say, any mistakes, omissions, and errors of interpretation that remain are entirely my responsibility. I also want to thank my literary agent, Peter Tallack, for his enthusiasm and energy, and my editor at University of Chicago Press, Christie Henry, for her enthusiasm for the idea, patience in its delivery, and unfailing encouragement. Finally I must thank my dear friend Jeremy Knight for reminding me of the rather dreadful old joke that I have included in the introduction!

SUGGESTIONS FOR FURTHER READING

GENERAL READING

Byers, Sean G., et al. "Natural Selection in a Contemporary Human Population." *PNAS* 107 (January 26, 2010): 1787–92. Plus other papers in this Sackler Colloquium Special Edition "Evolution in Health and Medicine."

Ewald, Paul. *Evolution of Infectious Disease*. Oxford: Oxford University Press, 1994.

———. *Plague Time: The New Germ Theory of Disease*. New York: Anchor Books, 2002.

Finch, Caleb. *The Biology of Human Longevity: Inflammation, Nutrition, and Aging in the Evolution of Lifespans*. Burlington, MA: Elsevier, 2007.

Gluckman, Peter, Alan Beedle, and Mark Hanson. *Principles of Evolutionary Medicine*. Oxford: Oxford University Press, 2009.

Gluckman, Peter, and Mark Hanson. *Mismatch: Why Our World No Longer Fits Our Bodies*. Oxford: Oxford University Press, 2006.

Henneberg, Maciej, and Arthur Saniotis. "How Can Evolutionary Medicine Inform Future Personalized Medicine?" *Personalized Medicine* 9 (2012): 171–73.

Nesse, Randolph M., and George C. Williams. "Evolution and the Origins of Disease." *Scientific American*, November 1998.

———. *Why We Get Sick: The New Science of Darwinian Medicine*. New York: Vintage, 1994.

Stearns, Stephen C., and Jacob C. Koella, eds. *Evolution in Health and Disease*. 2nd ed. Oxford: Oxford University Press, 2008.

Stearns, Stephen C., Randolphe M. Nesse, Diddahally R. Govindaraju, and Peter T. Ellison. "Evolutionary Perspectives on Health and Medicine." *PNAS* 107 (January 26, 2010): 1691–95.

Trevathan, Wenda R., E. O. Smith, James J. McKenna, eds. *Evolutionary Medicine*. 2nd ed. New York: Oxford University Press, 2007.

ABSENT FRIENDS

Ackerman, Jennifer. "The Ultimate Social Network." *Scientific American*, June 2012.

Blaser, Martin. *Missing Microbes: How Killing Bacteria Creates Modern Plagues*. New York: Oneworld/Henry Holt, 2014.

Eberl, Gerard. "A New Vision of Immunity: Homeostasis of the Superorganism." *Nature Mucosal Immunity*. May 5, 2010.

"Gut Microbes and Health." Special Section, *Nature Insight* 489 (September 13, 2012).

"The Gut Microbiota." Special Section, *Science Translational Medicine* 336 (June 8, 2012).

"Helminthic Therapy." http://opensourcehelminththerapy.org/mediawiki2/index.php?title =Main_Page.

Lee, Yun Kyung, and Mazmanian, Sarkis. "Has the Microbiota Played a Critical Role in the Evolution of the Adaptive Immune System?" *Science* 330 (December 24, 2010).

Mazmanian, Sarkis. Faculty home page. http://www.bbe.caltech.edu/content/sarkis -mazmanian.

"Recent Publications from the Weinstock Lab." http://sackler.tufts.edu/Faculty-and -Research/Faculty-Publications/Weinstock-Publications.

Rick Maizels' Group. Helminths and the Immune System. http://maizelsgroup.biology.ed .ac.uk/publications.

Rook, Graham A. W., ed. *The Hygiene Hypothesis and Darwinian Medicine*. Progress in Inflammation Research Series. London: Birkhauser, 2009.

A FINE ROMANCE

Brosens, Jan. "Further Publications." http://[illegible] =D17RnTsm77c=&inst=WARWICK.

Gangestad, Steven et al. "On the Function of Placental Corticotrophin-Releasing Hormone: A Role in Maternal-Fetal Conflicts over Blood Glucose Concentrations." *Biological Reviews* 87 (2012): 856–73.

Groopman, Jerome. "The Preeclampsia Puzzle." *New Yorker*. July 24, 2006.

Haig, David. "Genetic Conflicts in Human Pregnancy." *Quarterly Review of Biology* 68 (1993): 495–532.

———. "Troubled Sleep: Night Waking, Breastfeeding and Parent-Offspring Conflict." *Evolution, Medicine, and Public Health* March 7, 2014.

Robertson, Sarah A. "Immune Regulation of Conception and Embryo Implantation: All about Quality Control? *Journal of Reproductive Immunology* 85 (2010): 51–57.

Ubeda, Francisco, and David Haig. "Dividing the Child." *Genetica* 117 (January 2003): 103–10.

Wilkins, Jon F., and Haig, David. "What Good Is Genomic Imprinting: The Function of Parent-Specific Gene Expression." *Nature Reviews Genetics* 4 (May 2003).

Yuan, Hai-Tao, David Haig, and S. Ananth Karumanchi. "Angiogenic Factors in the Pathogenesis of Preeclampsia." *Current Topics in Developmental Biology* 71 (2005).

Zimmer, Carl. "Silent Struggle: A New Theory of Pregnancy." *New York Times*, March 14, 2006.

THE DOWNSIDE OF UPRIGHT

Ackerman, Jennifer. "The Downside of Upright." *National Geographic*, July 2006.
"Brains Plus Brawn: A Conversation with Daniel Lieberman." *Edge*. October 18, 2012.
Bramble, Dennis M., and Daniel E. Lieberman. "Endurance Running and the Evolution of *Homo*." *Nature* 432 (November 18, 2004).
Buccini, Cynthia K. "One Small Step for Man: Were The First Humans Walkers Or Tree Climbers? An Anthropologist Puts the Clues Together." *Bostonia*, Summer 2010.
Kivell, Tracy L., and Daniel Schmitt. "Independent Evolution of Knuckle-Walking in African Apes Shows that Humans Did Not Evolve from a Knuckle-Walking Ancestor." *PNAS* 106 (August 25, 2009): 14241–46.
Latimer, Bruce. "The Perils of Being Bipedal." *Annals of Biomedical Engineering* 33 (January 1, 2005).
Lieberman, Daniel E. "What We Can Learn about Running from Barefoot Running: An Evolutionary Medical Perspective." *Exercise and Sports Science Reviews* 40, no. 2 (2012).
Myerson, Julie. "Richard III, Scoliosis and Me." *Guardian*, February 5, 2013.
Whitcome, Katherine K., Liza J. Shapiro, and Daniel E. Lieberman. "Fetal Load and the Evolution of Lumbar Lordosis in Bipedal Hominins." *Nature* 450 (December 13, 2007).

DIY EYE

Akst, Jef. "Eyes Grown from Stem Cells." *Scientist*, April 6, 2011.
Eiraku, Mototsugu, et al. "Self-Organizing Optic-Cup Morphogenesis in Three-Dimensional Culture." *Nature* 472 (April 7, 2011).
Gehring, Walter J. "New Perspectives on Eye Development and the Evolution of Eyes and Photoreceptors." *Journal of Heredity* 93 (2005): 171–84.
Gollisch, Tim, and Markus Meister. "Eye Smarter than Scientists Believed: Neural Computations in Circuits of the Retina." *Neuron* 65 (January 28, 2010).
Kröger, Ronald H. H., and Oliver Biehlmaier. "Space-Saving Advantage of an Inverted Retina." *Vision Research* 49 (2009): 2318–21.
Lamb, Trevor D. "Evolution of the Eye." *Scientific American*, July 2011.
Land, Michael F., and Dan-Eric Nilsson. *Animal Eyes*. Oxford: Oxford University Press, 2012.
Provis, Jan M., et al. "Anatomy and Development of the Macula: Specialisation and the Vulnerability to Macular Degeneration." *Clinical and Experimental Optometry* 88, no. 5 (September 2005).
Singh, Mandeep S., et al. "Reversal of End-Stage Retinal Degeneration and Restoration of Visual Function by Photoreceptor Transplantation." *PNAS* 110, no. 3 (January 15, 2013): 1101–6.

HOPEFUL MONSTERS

Aktipis, C. Athena, et al. "Overlooking Evolution: A Systematic Analysis of Cancer Relapse and Therapeutic Resistance Research." *PLoS ONE* 6, no. 11 (November 2011).

Aktipis, C. Athena, and Randolph M. Nesse. "Evolutionary Foundations for Cancer Biology." *Evolutionary Applications* 6, no. 1 (January 2013).
Breivik, Jarle. "Don't Stop for Repairs in a War Zone: Darwinian Evolution Unites Genes and Environment in Cancer Development." *PNAS* 98 (May 8, 2001): 5379–81.
Caulin, Aleah F., and Carlo C. Maley. "Peto's Paradox: Evolution's Prescription for Cancer Prevention." *Trends in Ecology and Evolution* 26, no. 4 (April 2011).
Gatenby, Robert A., et al. "Adaptive Therapy." Cancer Research 69, no. 11 (June 1, 2009): 4894–903.
Greaves, Mel. *Cancer: The Evolutionary Legacy*. Oxford: Oxford University Press, 2000.
———. "Darwinian Medicine: A Case for Cancer." *Nature Reviews Cancer* 7 (March 2007): 213–21.
Greaves, Mel, and Carlo C. Maley. "Clonal Evolution in Cancer." *Nature* 481 (January 19, 2012): 306–13.
Hanahan, Douglas, and Robert A. Weinberg. "The Hallmarks of Cancer." *Cell* 100 (January 7, 2000): 57–70.
———. "Hallmarks of Cancer: The Next Generation." *Cell* 144 (March 4, 2011): 646–74.
Lengauer, Christoph, Kenneth W. Kinzler, and Bert Vogelstein. "Genetic Instabilities in Human Cancers." *Nature* 396 (December 17, 1998).
Li, Yilong, et al. "Constitutional and Somatic Rearrangement of Chromosome 21 in Acute Lymphoblastic Leukaemia." *Nature*, March 23, 2014.
Merlo, Lauren M. F., et al. "Cancer as an Evolutionary and Ecological Process." *Nature Reviews Cancer* 6 (December 2006): 924–35.
New Answers for Cancer. *Scientific American Classics Medicine Collection*, September 2012.
Rajagopalan, Harith, and Christoph Lengauer. "Aneuploidy and Cancer." *Nature* 432 (November 18, 2004).
Reid, Brian J. Oesophageal Cancer Publications. http://sharedresources.fhcrc.org/profile/reid-brian.
Reid, Brian J., Rumen Kostadinov, and Carlo C. Maley. "New Strategies in Barrett's Esophagus: Integrating Clonal Evolutionary Theory with Clinical Management." *Clinical Cancer Research* 17 (2011): 3512–19.
Rook, Graham A. W., and Angus Dalgleish. "Infection, Immuno-Regulation, and Cancer." *Immunological Reviews* 240 (2011): 141–59.
Roschke, Anna V., et al. "Chromosomal Instability Is Associated with Higher Expression of Genes Implicated in Epithelial-Mesenchymal Transition, Cancer Invasiveness, and Metastasis and with Lower Expression of Genes Involved in Cell Cycle Checkpoints, DNA Repair, and Chromatin Maintenance." *Neoplasia* 10 (2008): 1222–30.
Sottoriva, Andrea, et al. "Intratumor Heterogeneity in Human Glioblastoma Reflects Cancer Evolutionary Dynamics." *PNAS* 110, no. 10 (March 5, 2013): 4009–14.
Stephens, Philip J., et al. "Massive Genomic Rearrangement Acquired in a Single Catastrophic Event during Cancer Development." *Cell* 144 (January 7, 2011): 27–40.
Stratton, Michael R., Peter J. Campbell, and P. Andrew Futreal. "The Cancer Genome." *Nature* 458 (April 9, 2009).

A PROBLEM WITH THE PLUMBING

Brainerd, Elizabeth. "Efficient Fish Not Faint-Hearted." *Nature* 389 (September 18, 1997): 229.

Farmer, Colleen. "Did Lungs and the Intracardiac Shunt Evolve to Oxygenate the Heart in Vertebrates?" *Palaeobiology* 23, no. 3 (1997): 358–72.

Global Innovations. "Heart Attacks Are Not Just a 'Plumbing' Problem but a 'Whole System' Condition." September 15, 2012. https://tginnovations.wordpress.com/2012/09/15/heart-attacks-are-not-just-a-plumbing-problem-but-a-whole-system-condition/.

Hansson, Goran K. "Inflammation, Atherosclerosis, and Coronary Heart Disease." *New England Journal of Medicine* 352 (April 21, 2005): 1685.

Hansson, Goran K., and Andreas Hermansson. "The Immune System in Atherosclerosis." *Nature Immunology* 12, no. 3 (March 2011): 204.

Hansson, Goran K., and Peter Libby. "The Immune Response in Atherosclerosis: A Double-Edged Sword." *Nature Reviews Immunology* 6 (July 2006): 508.

Heart Cells Foundation, Stem Cell Clinical Trials. http://www.heartcellsfoundation.com/clinical-trials/.

Jha, Alok. "Cell Scientists Aim to Rebuild Hearts with Reprogrammed Tissue." *Guardian*, January 28, 2013.

Ketelhuth, D. F., et al. "T cell–Based Therapies for Atherosclerosis." *Current Pharmaceutical Design* 19, no. 33 (2013): 5850–58.

Libby, Peter, Paul M. Ridker, and Goran K. Hansson. "Progress and Challenges in Translating the Biology of Atherosclerosis." *Nature* 473 (May 19, 2011): 317.

Passier, Robert, et al. "Stem-Cell-Based Therapy and Lessons from the Heart." *Nature* 453 (May 15, 2008): 322.

Ridker, Paul M. Faculty home page. http://researchfaculty.brighamandwomens.org/BRIProfile.aspx?id=778.

Riley, Paul. "Plumbing the Heart." *Nature* 464 (March 25, 2010): 498.

Schulman, Ivonne H., and Joshua M. Hare. "Key Developments in Stem Cell Therapy in Cardiology." *Regenerative Medicine* 7 (2012): 7–24.

Segers, Vincent F. M., and Richard T. Lee. "Stem-Cell Therapy for Cardiac Disease." *Nature* 451(February 2008): 937.

Thompson, Randall C., et al., "Atherosclerosis across 4000 years of Human History: The Horus Study of Four Ancient Populations." *Lancet* 381, no. 9873 (2013): 1211–22.

THREE SCORE YEARS—AND THEN?

Ball, Melvyn J. "The Essential Lesion of Alzheimer's Disease: A Surprise in Retrospect." *Journal of Alzheimer's Disease* 9, supp. 3 (2006): 29–33.

Holmes, C., et al. "Proinflammatory Cytokines, Sickness Behaviour and Alzheimer's Disease." *Neurology* 77, no. 3 (July 19, 2011): 212–18.

Krstic, Dimitrije, and Irene Kneusel. "The Air-Bag Problem: A Potential Culprit for Bench-to-Bedside Translational Efforts: Relevance for Alzheimer's Disease." *Acta Neuropathalogica Communications* 1 (2013): 62.

Lee, Hyoung-gon, et al. "Amyloid-ß in Alzheimer's Disease: The Null versus the Alternate Hypothesis." *Journal of Pharmacology and Experimental Therapeutics* 321, no. 3 (2007).

McGreevey, Sue. "Alzheimer's-Associated Protein May Be Part of the Innate Immune System." *Harvard Gazette*, March 9, 2010.

Mohammadi, Dara. "Could Arthritis Drug Combat Alzheimer's?" *Observer*, November 17, 2013.

Parihar, Mordhwaj S., and Gregory J. Brewer. "Amyloid Beta as a Modulator of Synaptic Plasticity." *Journal of Alzheimer's Disease* 22 (January 1, 2010): 741–63.

Perry, V. Hugh, Col Cunningham, and Clive Holmes. "Systemic Infections and Inflammation Affect Chronic Neurodegeneration." *Nature Reviews Immunology* 7 (February 1, 2007): 161–67.